Question Bank in
Agricultural Microbiology

About the Author

The author Dr. Shankar Prasad Sha is pioneer researcher in metagenomics studies of Fermented Foods, and Beverages microbiology. He is expertise on Agricultural Microbiology, Soil Microbiology, PCR-DGGE and NGS techniques used for the microbial diversity studies. He is working on yeasts diversity of Amylolytic Starters of North-East India using culture-dependant and culture-independent molecular tools. He did his Masters in Microbiology from University of North Bengal and now pursuing his M.Phil- Ph.D in Microbiology from Sikkim Central University, Gangtok, Sikkim. He is a recipient of National Award "Rastriya Prativa Ratna Award 2017" for his academic and research excellence. He has qualified National Eligibility Test (NET) 2014-I and II.

Question Bank in

Agricultural Microbiology

Shankar Prasad Sha RA-Post-Doctoral Fellow

Department of Microbiology

Sikkim University

Gangtok, Sikkim - 737102

A Paperback Division of

NEW INDIA PUBLISHING AGENCY

101, Vikas Surya Plaza, CU Block, LSC Market
Pitam Pura, New Delhi 110 034, India
Phone: + 91 (11)27 34 17 17 Fax: + 91(11) 27 34 16 16
Email: info@nipabooks.com
Web: www.nipabooks.com

Feedback at feedbacks@nipabooks.com

ISBN : 978-93-87973-37-4

Composed, Designed and Printed in India

Preface

The book is primarily meant for students appearing for competitive examinations such as ARS-NET, ICAR-JRF, ICAR-SRF, GATE and viva voce in Agricultural Microbiology and Microbiology.

Agricultural Microbiology is one of the important components of the syllabus in ARS-NET, ICAR-JRF, ICAR-SRF exams. The main aim of this book is to help students to review their knowledge of Agricultural Microbiology acquired through textbooks and to understand the subject with logical reasoning.

Microbiology is the study of microorganisms that cannot be seen with the naked eye. These include bacteria, viruses, fungi, prions, archaea, protozoa and algae which are collectively known as 'microbes'. These microbes play significant roles in biogeochemical cycle, nutrient cycling, biomass decomposition, biodegradation/bio deterioration, food fermentation, food spoilage, and in Sewage treatment, Agriculture and Pharmaceuticals. There are lots of industrial applications of micro-organisms and their activities are vitally significant for various processes.

The book will be published in two volumes, this is the first volume that has seven chapters: chuapter 1- agriculture microbiology, chapter 2- microbiology of water, chapter 3-food microbiology, chapter 4-industrial microbiology, chapter-5-general microbiology, chapter-6- microbial growth and groth control and chapter 7-microbial physiology. In this book I have tried to give complete and latest information on various aspects of microbiology in objectives form with their proper answer.

I am very much grateful to New India Publishing Agency for publishing this book. I am very much hopeful that this book will be very much used by students of Agricultural Microbiology, General Microbiology, Biotechnology.

Shankar Prasad Sha
Department of Microbiology
Sikkim University

Acknowledgments

I would like to dedicate this book to my beloved mother late Smt. Shanti Devi and my father Shri Ram Prasad Sha, and my late grandfather. Shri Gorakhprasad Sha with the blessings of them and almighty God I wrote this book. I am thankful to Ms. Kriti Ghatani, Assistant Professor in Microbiology, Raiganj University and my research colleague in Sikkim University for constant support, encouragement and technical assistance in the preparation of this book. I am very much thankful to my all family members who always encouraged me and always be with me in all circumstances. I would like to express my gratitude to my research guide Professor (Dr.) J.P. Tamang for his encouragement and blessings. In this book we have tried to give complete and latest information on various aspects of food microbiology, keeping in mind that our undergraduate and postgraduate students of Microbiology, Food Technology and Biotechnology disciplines will be benefited. I would like to express my deep gratitude to my School teachers Shri K.K Singh, Shri Arun Jha, Shri Arun Keshri and my University teacher Dr. K.K Yadav, Dr Dhanirai Chettri and Prof. Mondal for their encouragements and blessings. I am thankful to my elder brothers Shri Uttam Prasad Sha, Shri Arjun Prasad Sha, Shri Bhim Prasad Sha, Shri Indrajeet Prasad Sha, my nephew Mr. Sumit Prasad Sha (B.Sc Microbiology student, Silliguri College) and Mr. Adharsh Prasad Sha, Mr. Aditya Prasad Sha, Mr. Dev Prasad Sha and my all niece Sweta, Niketa, Supriya, Nisha, Anjali, Niharika for their best wishes and support. I am thankful to my juniours Mukesh Pandey, Purrushottam Kumar, Santosh Kumar, and Vishnu Kumar for their best wishes. I am very much grateful to New India Publishing Agency for publishing this book in their reputed publication house.

Shankar Prasad Sha

Contents

1

Agriculture Microbiology

1. The scientist Dobereiner coined the term?

A. Associative Symbiosis B. Symbiosis
C. Obligatory Symbiosis D. Mutualistic Symbiosis

Ans.-A

2. Which of the following organism is the smallest?

A. Bacteria B. Virus
C. Fungi D. Yeast Ans.-B

3. *Azospirillum* can be culture in the media?

A. Semi-solid Malate B. NA
C. NB D. All of these Ans.-D

4. What is the optimum temperatue for nitrification?

A. 60 ^{0}C B. 50^{0}C
C. 80^{0}C D. 30^{0}C Ans.-D

5. What is the optimum temperature for ammonification?

A. 40-60 ^{0}C B. 40-50^{0}C
C. 40-80^{0}C D. 20-30^{0}C Ans.-D

6. Slow sand filters are used for filtration of which microbes?

A. E.coli B. Bacillus
C. Salmonella D. Giardia Ans.-D

7. Leaf nodules formed by which microbes?

A. E.coli and klebsiella B. Bacillus and klebsiella
C. Salmonella and klebsiella D. Klebsiella and Beijernikia

Ans.-D

8. Anaerobic decomposer of soil is?
   A. *E.coli* B. Bacillus
   C. Clostridium D. Klebsiella Ans.-C
9. Most persistent form of humus is?
   A. Humic acid B. Folic acid
   C. Nitric acid D. Nitrate Ans.-A
10. First antifungal agent is?
   A. Nystatin B. Tetracycline
   C. Penicillin D. Chloramfenical Ans.-A
11. Streptomycin discovered by?
   A. Waksman B. Domagk
   C. Fleming D. Paul Ehrlich Ans.-A
12. Who coined Chemotherapy word first?
   A. Waksman B. Domagk
   C. Fleming D. Paul Ehrlich Ans.-D
13. Which is the world's most infectious microbe?
   A. Wolbachia pipientis B. E coli
   C. Bacillus D. Streptococcus Ans.-A
14. Aerobic sulphur oxidisers are?
   A. Beggiotoa and Thiothrix B. Bacillus
   C. Yeast D. Micrococcus Ans.-A
15. Anaerobic sulphur oxidisers are?
   A. Beggiotoa and Thiothrix B. Bacillus
   C. Chlorobium and chromatiun D. Micrococcus Ans.-C
16. Cyano-bacteria are?
   A. Gram positive B. Gram negative
   C. Both D. None of all Ans.-B
17. E. Coli is?
   A. Gram positive B. Gram negative
   C. Both D. None of all Ans.-B
18. Winogradsky column used for study of?
   A. Sulpher metabolism B. Sulphite metabolism
   C. Nitrate metabolism D. None of all Ans.-A

19. Heterocyst is formed by?
   A. Cyanobateria B. E.coli
   C. Bacillus D. None of all Ans.-A
20. Nitrogen fixation is done by which gene?
   A. Nif gene B. Rep gene
   C. Vir gene D. None of all Ans.-A
21. Nostoc and anabaena are?
   A. Cyanobateria B. E.coli
   C. Bacillus D. None of all Ans.-A
22. Nostoc and anabaena do?
   A. Nitogen fixation B. Carbon fixation
   C. Sulphur fixation D. None of all Ans.-A
23. Azola is a water fern fix nitrogen with?
   A. Anabeana B. E.coli
   C. Bacillus D. None of all Ans.-A
24. The *Azosipirillum halopraeferans* fix nitrogen with?
   A. Paddy B. Wheat
   C. Kollar grasses D. Soyabean Ans.-C
25. The *Pseudomonas* species of bacteria fix nitrogen with
   A. Paddy B. Kollar grasses
   C. Soyabean D. Wheat Ans.-B
26. The Azotobacter nitrocaptans fix nitrogen with
   A. Sugarcane B. heat
   C. Paddy D. none of these Ans.-A
27. *Azosipirillum lipoferum fix nitrogen in*
   A. Miaze and cereals B. Grasses
   C. Paddy D. Wheat Ans.-A
28. *Azotobacter diazotrophicus* fix nitrogen in
   A. Kollar grasses B. Rice and Sugarcane
   C. Soyabean D. Cereal Ans.-B
29. *Rhizobium trifolli* is a :
   A. Free living nitrogen fixer B. Non symbiotic fixer
   C. Obligatory symbiotic fixer D. Symbiotic nitrogen fixer
   Ans.-D

30. Nitrogen fixation done by free living BGA are
A. *Fungi* B. *Bacillus*
C. *Pichia* D. *Nostoc, Anabaena* Ans.-D

31. Hetrocysts found in BGA are
A. *Oscillatoria* B. *Pichia*
C. *Candida* D. *Nostoc, Anabaena* Ans.-D

32. Which is non heterocystous BGA?
A. *Nostoc, Anabaena* B. *Pichia*
C. *Oscillatoria* D. *Candida* Ans.-C

33. Gram staining method was discovered by
A. Christian gram B. Alfred Gram
C. Robertcook D. Louis Pasteur Ans.-A

34. The first reported algal nitrogenase is obtained from which BGA?
A. Xenococcus B. Wollea
C. *Oscillatoria* D. Anabaena cylindrical Ans.-D

35. Hetrocyst has high level of which chemical?
A. Fumerate B. Lactone
C. Glutamine acetate D. Glutamine synthase (GS) Ans.-D

36. In Rhizobium the presence uptake hydrogenase in BGA is beneficial as it helps to enhance the?
A. Protein synthesis B. Fermentation
C. Reproduction D. Nitrogen fixation Ans.-D

37. Nitrogenase enzyme is more sensitive to which gas?
A. Nitrogen B. Carbon
C. Oxygen D. All of these Ans.-C

38. Nitrogen fixing BGA isolated from rice fields are?
A. *Bacillus* B. *Anabaena and Calothrix*
C. *protozoa* D. *Occillatoria* Ans.-B

39. Predator of BGA is?
A. Bacillus B. Yeast
C. Pseudomonas D. Chytridious fungi Ans.-D

40. Algal and fungal associations form?
A. Lichens B. Heterocysts
C. Hormogonia D. None of the above Ans.-A

41. The fungal partner of lichen is?
A. Lycobiont B. Mycobiont
C. Phycobiont D. Tetrabiont Ans.-B

42. The algal partner of lichen is?
A. Lycobiont B. Mycobiont
C. Phycobiont D. Tetrabiont Ans.-C

43. The root nodules produced in plants by?
A. *Frankia* B. Mycobiont
C. Phycobiont D. Tetrabiont Ans.-C

44. Which scientist first isolated the pure bacterial culture by serial dilution method?
A. Joseph Lister B. Robert Koch
C. Winogradsky D. Beijernick Ans.- A

45. Who are the pioneer scientists in soil microbiology?
A. Joseph Lister B. Robert Koch
C. Winogradsky & Beijernick D. Waksman Ans.-C

46. Who are the pioneer scientists in microbiology?
A. Joseph Lister
B. Robert Koch and Louis Pasteur
C. Winogradsky & Beijernick
D. Waksman Ans.- B

47. Who are the pioneer scientists studied ammonification of organic nitrogenous substances by soil microbes?
A. Joseph Lister B. Robert Koch
C. Lipman and Brown D. Waksman Ans.-C

48. Which scientist developed the direct soil examination technique?
A. Joseph Lister B. Robert Koch
C. Winogradsky & Beijernick D. Conn Ans.-D

49. Which scientist developed the contact slide technique?
A. Joseph Lister B. Robert Koch
C. Winogradsky & Beijernick D. Rossi Ans.-D

50. Who are the pioneer scientists studied Mycorrhiza?
A. Joseph Lister B. Robert Koch
C. Lipman and Brown D. Rayner and Mellin Ans.-D

51. Who is the pioneer scientist studied Phyllosphere?
A. Joseph Lister B. Robert Koch
C. Lipman and Brown D. Ruinen Ans.-D

52. Who is the pioneer scientist studied Chemo-Autotrophic bacterial photosynthesis?
A. Joseph Lister B. Robert Koch
C. Lipman and Brown D. Van Niel Ans.-D

53. Who is the pioneer scientist studied on transformation of iron bacteria?
A. Joseph Lister B. Robert Koch
C. Lipman and Brown D. Starkey Ans.-D

54. Who is the pioneer scientist studied on anaerobic respiration by methane bacteria?
A. Joseph Lister B. Robert Koch
C. Lipman and Brown D. Barker Ans.-D

55. Who is the pioneer scientist studied on root nodule bacteria?
A. Joseph Lister B. Robert Koch
C. Lipman and Brown D. Allen and Allen Ans.-D

56. Who are the pioneer scientists studied on colourful Algae and Microalgae?
A. Joseph Lister B. Robert Koch
C. Lipman and Brown D. Fritsch, Fogg & Stewart Ans.-D

57. Who is the pioneer scientist gave the theory of Spontaneous Generation?
A. Joseph Lister B. Robert Koch
C. Lipman and Brown D. Aristotle Ans.-D

58. Which scientist considered as father of Bacteriology?
A. Joseph Lister B. Robert Koch
C. Lipman and Brown D. Antony Van Leeuwenhoek Ans.-D

59. Who is the pioneer scientist isolated root nodule bacterium from root nodules of leguminous plants?
A. Joseph Lister B. Robert Koch
C. Lipman and Brown D. Beijernick Ans.-D

60. Root nodulation of *Comptonia Peregrina* formed by?
A. *Candia* B. *Pichia*
C. *Bacillus* D. *Frankia* Ans.-D

61. Azotobacterin contains?
A. *Candia* B. *Pichia*
C. *Bacillus* D. *Azotobacter* Ans.-D

62. Phosphobacterin contains?
A. *Candia*
B. *Pichia*
C. *Bacillus*
D. Phosphate solubilising Bacteria Ans.-D

63. Nitrogen fixing stem nodules on *Casuarina* spp.caused by?
A. *Candia* B. *Pichia*
C. *Bacillus* D. *Frankia* Ans.- D

64. Dommergues and his associates have discovered stem nodules of?
A. *Candia* *B.* *Pichia*
C. *Bacillus* *D.* *Sesbania rostrata* Ans.-D

65. What is the full form of PGPR?
A. Plant growth promoting Rhizobacteria
B. Plant growth promoting Rhizobium
C. Plant growth promoting Rhizo
D. Plant growth promoting Rhizospirillum Ans.- A

66. Examples of PGPR?
A. *Psedomonads & Rhizobium* B. *Pichia*
C. *E coli* D. *LAB* Ans.- A

67. Example of stem as well as root modulating bacteria?
A. *Psedomonads & Rhizobium* B. *Pichia*
C. *E coli* D. *Azotobacter Caulinodans*
Ans.-D

68. Examples of bacteria helping in bioleaching are?
A. *Psedomonads & Rhizobium* B. *Pichia*
C. *E coli* D. *Thibacillus & Ferrobacillus*
Ans.-D

69. The most abundance microbial population of soil is?
A. *Fungi* B. *Bacteria*
C. *E coli* D. *LAB* Ans.-B

70. Examples of nitrifying bacteria?

A. *Psedomonads & Rhizobium*

B. *Pichia*

C. *E coli*

D. *Nitrosommnas and Nitrococcus* Ans.-D

71. Examples of Nitrogen fixing bacteria?

A. *Azotobacter & Rhizobium*

B. *Pichia*

C. *E coli*

D. *Nitrosommnas and Nitrococcus* Ans.-A

72. Examples of denitrifying bacteria?

A. *Azotobacter & Rhizobium*

B. *Pichia*

C. *Pseudomonas dentrificans*

D. *Nitrosommnas and Nitrococcus* Ans.-C

73. Examples of Methane producing bacteria?

A. *Azotobacter & Rhizobium*

B. *Methanobacter & Methanococcus*

C. *Pseudomonas dentrificans*

D. *Nitrosommnas and Nitrococcus* Ans.-B

74. Examples of Iron reducing bacteria?

A. *Gallionella*

B. *Methanobacter & Methanococcus*

C. *Pseudomonas dentrificans*

D. *Nitrosommnas and Nitrococcus* Ans.-A

75. Examples of Sulpher reducing bacteria?

A. *Desulfovibrio desulfucans*

B. *Methanobacter & Methanococcus*

C. *Pseudomonas dentrificans*

D. *Nitrosommnas and Nitrococcus* Ans.-A

76. Examples of Sulpher oxidiser's bacteria?

A. *Desulfovibrio desulfucans*

B. *Methanobacter & Methanococcus*

C. *Beggiatoa and Thiothrix*

D. *Nitrosommnas and Nitrococcus* Ans.-C

77. Examples of carbon dioxide reducer bacteria?

A. *Desulfovibrio desulfucans*

B. *Methanobacter & Methanococcus*

C. *Beggiatoa and Thiothrix*

D. *Nitrosommnas and Nitrococcus* Ans.-B

78. Examples of thermophilic actinobacteria?

A. *Desulfovibrio desulfucans*

B. *Methanobacter & Methanococcus*

C. *Beggiatoa and Thiothrix*

D. *Thermoactionomyces & Streptomyces* Ans.-C

79. The phages attacking BGA are known as ?

A. *Cyanophages*

B. *Methanobacter & Methanococcus*

C. *Beggiatoa and Thiothrix*

D. *Thermoactionomyces & Streptomyces* Ans.-C

80. The phages attacking bacteria are known as?

A. *Bacteriophages*

B. *Methanobacter & Methanococcus*

C. *Beggiatoa and Thiothrix*

D. *Thermoactionomyces & Streptomyces*

Ans.-A

81. After bacteria the most abundance population of microbes in soil is?

A. *Fungi*

B. *Methanobacter & Methanococcus*

C. *Beggiatoa and Thiothrix*

D. *Thermoactionomyces & Streptomyces* Ans.-A

82. After bacteria the most abundance population of microbes in soil is?

A. *Fungi*

B. *Methanobacter & Methanococcus*

C. *Beggiatoa and Thiothrix*

D. *Thermoactionomyces & Streptomyces* Ans.-A

83. The phages attacking fungi are known as?
   A. *Mycophages*
   B. *Methanobacter & Methanococcus*
   C. *Beggiatoa and Thiothrix*
   D. *Thermoactionomyces & Streptomyces* Ans.-A

84. Industrial production of penicillin done by which microbe?
   A. *Penicillium Chrysogenum*
   B. *Methanobacter & Methanococcus*
   C. *Beggiatoa and Thiothrix*
   D. *Thermoactionomyces & Streptomyces* Ans.-A

85. Streak Plate method is used for the?
   A. Colony purification B. Cell seperation
   C. *Mitosis* D. *Cell differenciation* Ans.-D

86. Which culture independent method to study soil microbial diversity?
   A. PCR-DGGE B. PCR
   C. Replication D. HPLC Ans.-A

87. Rhizospere termed first coined by?
   A. *Hiltner* B. *Louis Pasteur*
   C. Winogradsky D. *Mayer* Ans.-A

88. Rhizospere effect is seen with bacteria in the range of ?
   A. 10-20 times B. 10-30 times
   C. 10-40 times D. 10-50 times Ans.-A

89. Microbial seed inoculants are?
   A. *Azotobater &Rhizobium* B. *Louis Pasteur*
   C. Winogradsky D. *Mayer* Ans.-A

90. Microbial seed inoculants are?
   A. *Azotobater & Rhizobium* B. *Louis Pasteur*
   C. Winogradsky D. *Mayer* Ans.-A

91. Example of biofertilisers are?
   A. *Azotobater & Rhizobium* B. *Louis Pasteur*
   C. Winogradsky D. *Mayer* Ans.-A

92. Example of Aquatic fern is?
   A. *Azotobater & Rhizobium* B. *Azolla*
   C. Pine D. *Mayer* Ans.-A

93. Example of bio controlling agents is?

A. *Bacillus & Pseudomonas* B. *Louis Pasteur*

C. Winogradsky D. *Mayer* Ans.-A

94. Phosphate solubilisation done by?

A. *Pseudomonas, Bacillus & Aspergillus*

B. *Louis Pasteur*

C. Winogradsky

D. *Azotobater & Rhizobium* Ans.-A

95. Copper transformations done by?

A. *Pseudomonas, Bacillus & Aspergillus*

B. *Louis Pasteur*

C. *Desulfovibrio, Ecoli, Clostridium*

D. *Azotobater & Rhizobium* Ans.- C

96. Leaf nodule formation done by?

A. *Pseudomonas, Bacillus & Aspergillus*

B. *Klebsiella*

C. *Desulfovibrio, Ecoli, Clostridium*

D. *Azotobater & Rhizobium* Ans.-B

97. Which genus of *Rhizobium* forms nodulation in peas?

A. *R.leguminosarium*

B. *R. phaseolii*

C. *Desulfovibrio, Ecoli, Clostridium*

D. *Azotobater & Rhizobium* Ans.-A

98. Which genus of *Rhizobium* forms nodulation in Beans?

A. *R. phaseolii*

B. *R.leguminosarium*

C. *Desulfovibrio, Ecoli, Clostridium*

D. *Azotobater & Rhizobium* Ans.-A

99. Which genus of *Rhizobium* forms nodulation in Clover group?

A. *R. phaseolii* B. *R.leguminosarium*

C. *R. trifolii* D. *Azotobater & Rhizobium*

Ans.-C

100. Which genus of *Rhizobium* forms nodulation in Alfalfa group?
A. *R. phaseolii* B. *R.leguminosarium*
C. *R. trifolii* D. *R. meliloti* Ans.-D

101. Which genus of *Rhizobium* forms nodulation in Lupini group?
A. *R. lupini* B. *R.leguminosarium*
C. *R. trifolii* D. *R. meliloti* Ans.-A

102. Which genus of *Rhizobium* forms nodulation in Soyabean group?
A. *R. japonicum* B. *R.leguminosarium*
C. *R. trifolii* D. *R. meliloti* Ans.-D

103. Which genus of *Rhizobium* forms nodulation in Cowpea group?
A. *R. spp.* B. *R.leguminosarium*
C. *R. trifolii* D. *R. meliloti* Ans.-D

104. The genus of Sino*rhizobium* forms nodulation in which group of plants?
A. *Soyabean* B. Cowpea
C. Peas D. Lupin Ans.-A

105. The genus of *Azorhizobium* forms nodulation in which group of plants?
A. *Soyabean* B. Sesbabia
C. Peas D. Lupin Ans.-B

106. The QUANTUM contains which microbe?
A. *E. coli* B. *Bacillus subtilis*
C. *Pichia* D. *Candida* Ans.-B

107. The QUANTUM contains which microbe?
A. *E. coli* B. *Bacillus subtilis*
C. *Pichia* D. *Candida* Ans.-B

108. The potato plants producing which Phytoalexins?
A. *Chlorogenic Acid* B. *Pisatin*
C. *Phaseoline* D. Wyerone Ans.-A

109. The Pea plants producing which Phytoalexins?
A. *Chlorogenic Acid* B. *Pisatin*
C. *Phaseoline* D. Wyerone Ans.- B

109. The Green beans producing which Phytoalexins?
A. *Chlorogenic Acid* B. *Pisatin*
C. *Phaseoline* D. Wyerone Ans.-C

110. The Carrot plants producing which Phytoalexins?
A. *Chlorogenic Acid* B. *Pisatin*
C. *Phaseoline* D. Wyerone Ans.-A

111. The Soyabean plants producing which Phytoalexins?
A. *Chlorogenic Acid* B. *Pisatin*
C. Hydroxy *Phaseoline* D. Wyerone Ans.-C

112. The Sweet potato producing which Phytoalexins?
A. *Chlorogenic Acid* B. *Pisatin*
C. Hydroxy *Phaseoline* D. Wyerone Ans.-A

113. Cotton plants producing which Phytoalexins?
A. *Chlorogenic Acid* B. *Pisatin*
C. Hydroxy *Phaseoline* D. Gossypol Ans.-D

114. The Aflalfa plants producing which Phytoalexins?
A. *Chlorogenic Acid* B. *Pisatin*
C. Hydroxy *Phaseoline* D. Sativol Ans.-D

115. Tobacco plants producing which Phytoalexins?
A. *Chlorogenic Acid* B. *Pisatin*
C. Hydroxy *Phaseoline* D. Scopoletin Ans.-D

116. The apple plants producing which Phytoalexins?
A. *Chlorogenic Acid* B. *Pisatin*
C Hydroxy *Phaseoline* D. Phloridzin Ans.-D

117. The Broad beans producing which Phytoalexins?
A. *Chlorogenic Acid* B. *Pisatin*
C. Hydroxy *Phaseoline* D. Wyerone Ans.-D

118. The Flavodoxin is isolated from which bacteria?
A. *Clostridium pasteurianum* B. *Pichia*
C. *Candia* D. Bacillus Ans.-A

119. Which media is used for Flavodoxin is isolated from which bacteria?
A. *Clostridium pasteurianum* B. *Pichia*
C. *Candia* D. Bacillus Ans.-A

120. *Rhizobium* cells are?
A. *Gram negative rod* B. *Gram positive*
C. *Cocci* D. Gram positive rod Ans.-A

121. *Rhizobium* cells have what kind of flagella are?
A. Peritrichous flagella B. Monotrichous flagella
C. Monotrichous flagella D. Gram positive rod Ans.-A

122. *Rhizobium* cells grow on media?
A. YEMA B. NA
C. YMA D. NB Ans.-A

122. Which Flavonoids induce nodulation *in Alfalfa Plant?*
A. Luteonin B. Geraldon
C. Genistein D. Naringenin Ans.-A

123. Which Flavonoids induce nodulation *in clover Plant?*
A. Luteonin B. Geraldon
C. Genistein D. Naringenin Ans.-B

124. Which Flavonoids induce nodulation *in Soyabean Plant?*
A. Luteonin B. Geraldon
C. Genistein D. Naringenin Ans.-C

124. Which Flavonoids induce nodulation *in Beans Plant?*
A. Luteonin B. Geraldon
C. Genistein D. Delphinidin Ans.-B

125. Which Flavonoids induce nodulation *in clover Plant?*
A. Luteonin B. Geraldon
C. Genistein D. Naringenin Ans.-B

126. Which Flavonoids induce nodulation *in Vetch root Plant?*
A. Luteonin B. Geraldon
C. Genistein D. Naringenin Ans.-D

127. Which gene induces nodulation process?
A. NOD gene B. nif gene
C. Rep gene D. R gene Ans.-B

128. Indol Acetic Acid (IAA) produced by *Rhizobia* helps in?
A. Root hair curling and root nodulation
B. Metabolism
C. Phosphorylation
D. Evaporation Ans.- A

129. Plant protein Lectins helps in?
A. binding of *Rhizobia* with plant root
B. Metabolism
C. Phosphorylation
D. Evaporation Ans.-A

130. Trifolin protein synthesized by?
A. *R. japonicum* B. *R. leguminosarium*
C. *R. trifolii* D. *R. meliloti* Ans.-C

131. Stem modulating legumes are?
A. Sesbania and Aeschyniomene B. Potao
C. Soyabean D. Mango Ans.-A

132. Rhizobiotoxin produced by?
A. *R. japonicum* B. *R. leguminosarium*
C. *R. trifolii* D. *R. meliloti* Ans.-A

133. *Actino-rhizo* symbiosis caused by?
A. *Frankia* B. *R. leguminosarium*
C. *R. trifolii* D. *R. meliloti* Ans.-A

134. Callaham and his associates first reported the *Frankia* isolated from which plants?
A. *Comptonia peregrina* B. *Cotton*
C. *Potato* D. *Soyabean* Ans.-A

135. *Frankia* is?
A. Microaerophilic in nature B. Aerobic
C. Anareobic D. obligatory aeriobic Ans.-A

136. *Frankia* is slow growing actinomycetes having log phase of?
A. 14 days B. 7 days
C. 5 days D. 40 days Ans.-A

137. Plant residues contain what percentage of cellulose?
A. 15-60% B. 15-50%
C. 15-80% D. 15-90 Ans.-A

138. Plant residues contain what percentage of hemi-cellulose?
A. 10-30% B. 15-50%
C. 15-80% D. 15-90 Ans.-A

139. Plant residues contain what percentage of lignin?
A. 5-30 % B. 15-50%
C. 15-80% D. 15-90 Ans.-A

140. Plant residues contain what percentage of protein?
A. 2-15% B. 5-50%
C. 15-80% D. 15-90 Ans.-A

141. Plant residues contain what percentage of sugars?
A. 10% B. 15-50%
C. 15-80% D. 15-90 Ans.-A

142. Pectine is an example of hemicelluloses degraded by?
A. Enzyme Pectinase B. lipase
C. amylase D. Glucoamylase Ans.-A

143. Lignin is degraded by?
A. *Clitocybe,collybia* B. *Trichoderma*
C. *Pichia* D. *Bacillus* Ans.-A

144. Starch is degraded by fungus?
A. *Aspergillus,Rhizopus* B. *Trichoderma*
C. *Pichia* D. *Bacillus* Ans.-A

145. Pectine is degraded by?
A. Fusarium and Verticillium B. *Trichoderma*
C. *Pichia* D. *Bacillus* Ans.-A

146. Tanin is degraded by?
A. *Aspergillus & Penicillium* B. *Trichoderma*
C. *Pichia* D. *Bacillus* Ans.-A

147. Fulvic acid is utilised by?
A. *Poria* B. *Trichoderma*
C. *Pichia* D. *Bacillus* Ans.-A

148. Cutin is utilised by?
A. *Aspergillus & Rhodotorulla* B. *Trichoderma*
C. *Pichia* D. *Bacillus* Ans.-A

149. Humic acid is utilised by?
A. *Penicillium and Polysticus* B. *Trichoderma*
C. *Pichia* D. *Bacillus* Ans.-A

150. Hemicellulose degraded by which fungus?
A. *Penicillium & Aspergillus* B. *Trichoderma*
C. *Pichia* D. *Bacillus* Ans.-A

151. *Hemicellulose* degraded by which Bacteria?
A. *Bacillus & Pseudomonas* B. *Trichoderma*
C. *Pichia* D. *Bacillus* Ans.-A

152. Cellulose degraded by which Bacteria?
A. *Bacillus & Pseudomonas* B. *Trichoderma*
C. *Pichia* D. *Bacillus* Ans.-A

153. Cellulose degraded by which fungus?
A. *Penicillius, Aspergillus, Trichoderma*
B. *Trichoderma*
C. *Pichia*
D. *Bacillus* Ans.-A

154. Starch degraded by which bacteria?
A. *Bacillus and Clostridium* B. *E.coli*
C. *Serretia* D. *Micrococcus* Ans.-A

155. Chitin degraded by which bacteria?
A. *Bacillus and Clostridium* B. *E.coli*
C. *Serretia* D. *Micrococcus* Ans.-A

156. Protein degraded by which bacteria?
A. *Bacillus, Pseudomonas, Clostridium*
B. *E.coli*
C. *Serretia*
D. *Micrococcus* Ans.-A

157. Fulvic acid is?
A. alkali as well as acid soluble
B. Acid non-solube
C. Alkali soluble
D. Base soluble Ans.-A

158. Humin is acid is resistant to?
A. Cold alkali B. Base
C. Alkali soluble D. Base soluble Ans.-A

159. Humic acid are regarded as polymers of?

A. Aeromatic compounds B. Base

C. Alkali soluble D. Base soluble Ans.-A

160. Which family of fungi mostly able to degrade the lignin?

A. Basidiomycetes & Ascomycetes B. Phycomycetes

C. Deuteromycetes D. Non-Deuteromycetes

Ans.-A

161. Benefits of Humic acid are?

A. Seed germination, root growth, uptake of nutrients

B. Nodulation

C. Replication

D. decomposition Ans.-A

162. EDTA is well known?

A. Synthetic Chelator

B. Protein

C. Carbohydrate

D. Synthetic antibiotic Ans.-A

163. What is the optimum C/N ration is ideal for the decomposition of organic matters?

A. C/N (20-25%) B. C/N (20-50%)

C. C/N (60-70%) D. C/N (30-60%) Ans.-A

164. Mineralization is the process of?

A. Degradation of organic matters

B. Uptake of organic matters

C. Synthesis of organic matters

D. Distribution of Organic matters Ans.-A

165. Immobilization is the process of?

A. Degradation of organic matters

B. Uptake of organic matters by microbes

C. Synthesis of organic matters

D. Distribution of Organic matters Ans.-B

166. Vermicompost is the process of?

A. Use of earthworm in composting

B. Uptake of organic matters by microbes

C. Synthesis of organic matters
D. Distribution of Organic matters Ans.-A

167. Vermicompost enhances soil fertility by increasing which nutrient?
A. Calcium B. Potassium
C. Zinc D. Sulphur Ans.-A

168. What is the optimum C/N ratio is ideal for the best composting?
A. C/N (15-20%) B. C/N (20-50%)
C. C/N (60-70%) D. C/N (30-60%) Ans.-A

169. What is the optimum organic matter content for the best composting?
A. (30-60%) B. (20-50%)
C. (60-70%) D. (30-60%) Ans.-A

170. A composting is considered superior at what ratio of C/N?
A. C/N (20%) B. C/N (20-50%)
C. C/N (60-70%) D. C/N (30-60%) Ans.-A

171. Which is the thermo-tolerant earthworm helping in Vermicomposting?
A. *Lumbricus rubellusa* and *Eisenia foetida*
B. *Streptomyces*
C. *Nocardia*
D. *Frankia* Ans.-A

172. Example of Green manure?
A. *Sesbani rostrata* B. *Streptomyces*
C. *Nocardia* D. *Frankia* Ans.-A

173. What is the methane contains of Biogas?
A. (50-60%) B. (20-50%)
C. (60-70%) D. (30-60%) Ans.-A

174. What is the hydrogen contains of Biogas?
A. (5-10%) B. (20-50%)
C. (60-70%) D. (30-60%) Ans.-A

175. What is the carbon dioxide contains of Biogas?
A. (30-35%) B. (20-50%)
C. (60-70%) D. (30-60%) Ans.-A

176. What is the major substrate of Biogas?
A. Cow-dung B. Wood
C. Coal D. Oil Ans.-A

177. What is the optimum temperature for the Biogas production?

A. 30° B. 20°

C. 50° D. 80° Ans.-A

178. Which bacteria plays important role in Biogas production?

A. Methane bacteria B. Hydrogen Bacteria

C. *Bacillus* D. *Rhizobium* Ans.-A

179. Ethan, a short chain hydrocarbon is degraded by which microbes?

A. *Nocardi, Streptomyces, Pseudomonas*

B. Hydrogen Bacteria

C. *Bacillus*

D. *Rhizobium* Ans.-A

180. Ethan, a short chain hydrocarbon is degraded by which bacteria?

A. *Pseudomonas & Flavobaterium*

B. Hydrogen Bacteria

C. *Bacillus*

D. *Rhizobium* Ans.-A

181. Ethan, a short chain hydrocarbon is degraded by which Actinomycetes?

A. *Nocardia & Streptomyces* B. Hydrogen Bacteria

C. *Bacillus* D. *Rhizobium* Ans.-A

182. High molecular weight hydrocarbon is degraded by which microbes?

A. *Nocardia, Streptomyces, Pseudomonas*

B. Hydrogen Bacteria

C. *Bacillus*

D. *Rhizobium* Ans.-A

183. Gibberelic Acid is produced by?

A. *Gibberella fujikuroi* B. Hydrogen Bacteria

C. *Bacillus* D. *Rhizobium* Ans.-A

184. Gibberelic Acid is produced by which fungi?

A. *Penicillium & Rhizopous* B. Hydrogen Bacteria

C. *Bacillus* D. *Rhizobium* Ans.-A

185. Cytokinins are produced by?

A. *Azotobacter & Azospirillum* B. Hydrogen Bacteria

C. *Bacillus* D. *Rhizobium* Ans.-A

186. Ethylene is produced by?
A. *Citrobacter & Pseudmonas* B. Hydrogen Bacteria
C. *Bacillus* D. *Rhizobium* Ans.-A

187. Penicillin is produced by?
A. *Penicillim notatum* B. Hydrogen Bacteria
C. *Bacillus* D. *Rhizobium* Ans.-A

188. Amphotericin B is produced by?
A. *Streptomyces nodosus* B. Hydrogen Bacteria
C. *Bacillus* D. *Rhizobium* Ans.-A

189. Bacitracin is produced by?
A. *Bacillus subtilis* B. Hydrogen Bacteria
C. *Bacillus* D. *Rhizobium* Ans.-A

190. Fusidic acid is produced by?
A. *Fusarium coccineum* B. Hydrogen Bacteria
C. *Bacillus* D. *Rhizobium* Ans.-A

191. Erythromycin is produced by?
A. *Streptomyces* B. Hydrogen Bacteria
C. *Bacillus* D. *Rhizobium* Ans.-A

192. Neomycin is produced by?
A. *Streptomyces* B. Hydrogen Bacteria
C. *Bacillus* D. *Rhizobium* Ans.-A

193. Nystatin is produced by?
A. *Streptomyces spp.* B. Hydrogen Bacteria
C. *Bacillus* D. *Rhizobium* Ans.-A

194. Kanamycin is produced by?
A. *Streptomyces spp.* B. Hydrogen Bacteria
C. *Bacillus* D. *Rhizobium* Ans.-A

195. Choramphenical is produced by?
A. *Streptomyces venezuelae* B. Hydrogen Bacteria
C. *Bacillus* D. *Rhizobium* Ans.-A

196. Polymyxin B is produced by?
A. *Aerobacillus polymyxa* B. Hydrogen Bacteria
C. *Bacillus* D. *Rhizobium* Ans.-A

197. Viomycin is produced by?

A. *Streptomyces* spp. B. Hydrogen Bacteria

C. *Bacillus* D. *Rhizobium* Ans.-A

198. Spiramycin is produced by?

A. *Streptomyces* spp. B. Hydrogen Bacteria

C. *Bacillus* D. *Rhizobium* Ans.-A

199. Gentamycin is produced by?

A. *Micromonospora purpurea* B. Hydrogen Bacteria

C. *Bacillus* D. *Rhizobium* Ans.-A

200. Vancomycin is produced by?

A. *Streptomyces orientalis* B. Hydrogen Bacteria

C. *Bacillus* D. *Rhizobium* Ans.-A

2

Microbiology of Water

1. The concept of putting microbes to help clean up the environment is called

A. Pasteurization B. Bioremediation

C. Fermentation D. Biolistics Ans.-B

2. Which of the following is not employed as an oxidation method?

A. Oxidation ponds B. Trickling filters

C. Contact aerators D. All of these Ans.-D

3. The filtering medium of trickling filters is coated with microbial flora, known as

A. Zoological film B. Geological film

C. Zooglocal film D. None of these Ans.-C

4. The biogas production process takes place at the temperature

A. Lesser than 25°C B. 25-40°C

C. 45-60°C D. All of these Ans.-D

5. Advanced treatment is generally used to treat waste water to

A. Remove coarse solids

B. Remove settleable solids

C. Reduce BOD

D. Remove additional objectionable substances Ans.-D

6. What is an anaerobic digester?

A. New diet drink

B. Microbe that eats hazardous waste

C. Method to convert agricultural waste into a biogas

D. All of the above Ans.-A

7. The use of microbes to breakdown synthetic waste products such as polychlorinated biphenyls is called

A. Bioinformatics B. Biolistics

C. Biotechnology D. Bioremediation Ans.-C

8. Activated sludge contains large number of

A. Bacteria B. Yeasts and molds

C. Protozoa D. All of these Ans.-D

9. Iron bacteria can produce

A. Slime B. Undesirable odors and tastes

C. Both (a) and (b) D. Extreme acidity Ans.-C

10. The microbiological examination of coliform bacteria in foods preferably use

A. MacConkey broth B. Violet Red Bile agar

C. Eosine Methylene blue agar D. All of these Ans.-D

11. Which of the following acid will have higher bacteriostatic effect at a given pH?

A. Acetic acid B. Tartaric acid

C. Citric acid D. Maleic acid Ans.-A

12. Water activity can act as

A. An intrinsic factor determining the likelihood of microbial proliferation

B. A processing factor

C. An extrinsic factor

D. All of the above Ans.-???

13. What are the intrinsic factors for the microbial growth?

A. pH B. Moisture

C. Oxidation-Reduction Potential D. All of these Ans.-D

14. Yeast and mould count determination requires-

A. Nutrient agar

B. Acidified potato glucose agar

C. MacConkey agar

D. Violet Red Bile agar Ans.-B

15. The time temperature combination for HTST Paterurization of 71.1°C for 15 sec is selected on the basis of

A. *Coxiella Burnetii* B. *E. coli*

C. *B. subtilis* D. *C. botulinum* Ans.-A

16. Aerobic Colony Count (ACC), is also known as
    A. Total viable count (TVC) B. Aerobic plate count (APC)
    C. Standard plate count (SPC) D. All of these Ans.-D
17. Suspected colonies of *Staphylococcus aureus* when grown on Baird-Parker medium shall show
    A. coagulase activity B. protease activity
    C. catalase activity D. none of these Ans.-A
18. A psychrophilic halophile would be a microbe that prefers
    A. cold temperatures and increased amounts of salt
    B. warm temperatures and increased amounts of pressure
    C. cold temperatures and the absence of oxygen
    D. warm temperatures and increased amounts of acid Ans.-A
19. NaCl can act as
    A. antagonist at optimal concentrations
    B. synergistically if added in excess of optimum level
    C. Both (a) and (b)
    D. None of the above Ans.-D
20. The MPN and membrane fitration techniques are used for the detection of?
    A. *Coliform Detection* B. Bacillus detection
    C. Pichia D. Candida Ans.-A
21. Coliform consists of which bacteria?
    A. *E. coli, Sheigella, salmonella* B. Bacillus detection
    C. Pichia D. Candida Ans.-A
22. Indicator organisms are?
    A. *E. coli, Sheigella, salmonella* B. Bacillus detection
    C. Pichia D. Candida Ans.-A
22. Index organisms are?
    A. *E. coli, Sheigella, salmonella* B. Bacillus detection
    C. Pichia D. Candida Ans.-A
23. 2, 4, 6 Trinitro Toluene degraded by ?
    A. *Pseudomonas and Clostridium* spp.
    B. Bacillus detection
    C. Pichia D. Candida Ans.-A

24. Acid forming bacteria are?
    A. Clostridium, Lactobacillus, *E.coli*
    B. Bacillus
    C. Pichia
    D. Candida — Ans.-A

25. Methanogenic bacteria are?
    A. Methaobacter, Methanococcus B. Bacillus
    C. Pichia D. Candida — Ans.-A

26. Wetland used for?
    A. Denitrification and metal removal
    B. Nitrification
    C. Ammonification
    D. nitrogen Fixation — Ans.-A

27. *Zoogloea ramigera* used for?
    A. Floc formation during sewage treatment
    B. Nitrification
    C. Ammonification
    D. nitrogen Fixation — Ans.-A

28. For water purification how much chlorine used?
    A. 0.2 to 2.0 mg/lt B. 0.2 to 1.0 mg/lt
    C. 0.2 to 6.0 mg/lt *D.* 0.2 to 3.0 mg/lt — Ans.-A

29. Ozonization is used for?
    A. *Water purification* B. Air purification
    C. *Soil purification* *D.* *Oil purification* — Ans.-A

30. Water pathogens are?
    A. *Vibrio cholera, Legionella* B. *E.coli*
    C. *Phichia* D. *Candida* — Ans.-A

31. Cysts of *Giadia* filtered through?
    A. *Slow sand filter* B. *Filter*
    C. *Sand filter* D. *Microfilter* — Ans.-A

32. Viruses are inactivated by?
    A. Chlorine, bromine, formaldehide
    B. Filter

C. Sand filter
D. Microfilter Ans.-A

33. MPN indicates the quality of?
A. Potabiliy of Water B. Potabiliy of Air
C. Potabiliy of Soil D. Potabiliy of Oil Ans.-A

34. Eutrophication is?
A. Nutrient rich condition B. Nutrient less condition
C. Solubility D. Non solubility Ans.-A

35. Oligotrophic condition is?
A. Nutrient rich condition B. Nutrient deficient condition
C. Solubility D. Non solubility Ans.-B

36. The examples of Phytoplankton are?
A. Algae B. Protozoa
C. Bacteria D. Fungi Ans.-A

36. The examples of Zooplantons are?
A. Algae B. Protozoa
C. Bacteria D. Fungi Ans.-B

37. Barophilic Microbes are?
A. Presure loving B. Heat loving
C. Acid loving D. Salt loving Ans.-A

38. The test perfomed to distinguisg between *E.coli* and Enterobacter .aerogens is?
A. IMVIC test B. VVIC test
C. Chlorine test D. TNIC test Ans.-A

39. In membrane fitration test the microbes are retained on?
A. On surface of membrane filter
B. Inside surface of membrane filter
C. In water
D. In air Ans.-A

40. Water of good quality should have less than how many bacterial colonies?
A. Less than 100 B. More than 100
C. 10000 D. 20000 Ans.-A

41. In municipal water supplies the main water disinfectant is?
A. Chlorine B. Iodine
C. Bromine D. Sunlight Ans.-A

42. Sewage contains which bacteria?
A. *Sphaerotilus* B. *E.coli*
C. *Bacillus* D. *Pseudomonas* Ans.-A

43. Sewage contains which fungi?
A. *Saprolegina* B. *Pichia*
C. *Candida* D. *Trichoderma* Ans.-A

44. Zoogloial film produced on?
A. Trickling Filter B. Oxidation pond
C. Wetland D. River Ans.-A

45. Trickling Filter is prepared in which phase of water treatment?
A. During Secondary treatment B. Primary treatment
C. Tertiary treatmet D. None Ans.-A

46. Oxidation ponds or lagoons are helpful in the process of?
A. Oxidation of organic matters B. Primary treatment
C. Tertiary treatmet D. None Ans.-A

47. Activated sludge process is involed during which phase of sewage treatment?
A. Oxidation of organic matters B. Primary treatment
C. Tertiary treatmet D. None Ans.-A

48. BOD high means?
A. More pollution in water B. More pollution in air
C. More pollution in soil D. None Ans.-A

49. BOD high then oxygen demand by microbes in water will be?
A. Low B. High
C. Midium D. Very Low Ans.-B

48. BOD high means?
A. More pollution in water B. More pollution in air
C. More pollution in soil D. None Ans.-A

50. Sea water contains chlorine?
A. 19.4g/kg B. 17.4g/kg
C. 11.4g/kg D. None of all. Ans.-A

51. What is bioaggumentation?
A. Addition of bacteria cells inside soil B. Removal of bacterial cells
C. Addition of plants D. Addition of chemicals Ans.-A

52. What is bioremediation?
A. Detoxification of Pollutants B. Removal of bacterial cells
C. Addition of plants D. Addition of chemicals Ans.-A

53. What is benthic zone?
A. Bottom of river B. Top of river
C. Top of sea D. River side Ans.-A

53. What is lotic condition?
A. Running water B. Stangnent water
C. Stable water D. River side Ans.-A

54. What is lantic condition?
A. Running water B. Stangnent water
C. Stable water D. River side Ans.-B

55. What is O horizon?
A. Upper layer of soil B. Stangnent water
C. Stable water D. River side Ans.-A

56. What is C horizon?
A. Lower layer of soil B. Stangnent water
C. Stable water D. River side Ans.-A

57. Microbial activity is high in horizon?
A. A horizon B. B horizon
C. O horizon D. C horizon Ans.-A

58. The primary producers are?
A. Plants B. Animals
C. Protozoa D. Fungi Ans.-A

59. The primary producers use light for?
A. Photosynthesis B. Respiration
C. Fermentation D. Transpiration Ans.-A

60. The prim ary producers do?
A. Photosynthesis B. Respiration
C. Fermentation D. Transpiration Ans.-A

61. Hydrocarbon degrading bacteria are?
A. *Pseudomonas* B. *E coli*
C. Shiegella D. Vibrio Ans.-A

62. Hydrocarbon degrading bacteria are?
A. *Pseudomonas* B. *E coli*
C. Shiegella D. Vibrio Ans.-A

63. Oil degrading bacteria are?
A. *Alkalnivorax borjumensis* B. *E coli*
C. Shiegella D. Vibrio Ans.-A

64. Petrolium degrading bacteria are?
A. *Pseudomonas* B. *E coli*
C. Shiegella D. Vibrio Ans.-A

65. Example of xenobiotic compounds are?
A. *DDT, PCBS* B. *E coli*
C. Shiegella D. Vibrio Ans.-A

66. Example of petroleum producing algae is?
A. *B.braunii* B. *E coli*
C. Shiegella D. Vibrio Ans.-A

67. Example of microbe who do aerobic dechorination?
A. *Pseudomonas and Burkholderia*
B. E coli
C. Shiegella
D. Vibrio Ans.-A

68. Example of rumen bacteria?
A. *Ruminococcus albus* B. *E coli*
C. Shiegella D. Vibrio Ans.-A

69. Example of cellulose decomposer in rumen?
A. *Ruminococcus albus* B. *E coli*
C. Shiegella D. Vibrio Ans.-A

70. Example of starch decomposer in rumen?
A. *Succinomonas amylolytica* B. *E coli*
C. Shiegella D. Vibrio Ans.-A

71. Example of Lactate decomposer in rumen?

A. *Selenomonas lactilytica* B. *E coli*

C. Shiegella D. Vibrio Ans.-A

72. Example of succinate decomposer in rumen?

A. *Schwartzia succinovorans* B. *E coli*

C. Shiegella D. Vibrio Ans.-A

73. Example of Pectin decomposer in rumen?

A. *Lachnospirab multiparus* B. *E coli*

C. Shiegella D. Vibrio Ans.-A

74. Example of methanogens in rumen?

A. *Methanobrevibacter ruminantium*

B. *E coli*

C. Shiegella

D. Vibrio Ans.-A

75. The examples bioluminescence bacteria 0020?

A. Vibrio harveyi B. *E coli*

C. Shiegella D. Vibrio Ans.-A

3

Food Microbiology

1. Yogurt fermented with help of what kind of bacteria?
   A. *Lactobacillus bulgaricus & Streptococcus thermophillus*
   B. L.acidophilus
   C. Lcatococcus
   D. *Citrrobacter* Ans.-A
2. Vinegar fermentetation done with help of what kind of bacteria?
   A. Lactic acid bacteria B. Acetic acid bacteria
   C. Formic acid bacteria D. Citric acid bacteria Ans.-B
3. Vinegar fermentetation done with help of which method?
   A. Trickling method B. Trickle method
   C. Fricling method D. Filtering method Ans.-B
4. Soya sauce fermentation done with help which fungus?
   A. *Aspergillus oryzae & A. soyae*B. Penicillium
   C. Rhizopous D. Mucor Ans.-A
5. Yogurt is produced with help what kind of bacteria?
   A. *Lactobacillus* sp. B. Acetic acid bacteria
   C. Formic acid bacteria D. Citric acid bacteria Ans.-A
6. Sour cream obtained by?
   A. *Streptococcus lactis & L.citrovorum*
   B. Aspergillus oryzae & A. soyae
   C. Formic acid bacteria
   D. Citric acid bacteria Ans.-A

7. Rice beer is produced with help which yeast?

A. Saccharomyces | B. Candida
C. Pichia | D. Issatchenkia | Ans.-A

8. Which is known as brewers yeast?

A. Saccharomyces cerevisiae | B. Candida
C. Pichia | D. Issatchenkia | Ans.-A

9. Which is known as distillers yeast?

A. Saccharomyces cerevisiae | B. Candida
C. Pichia | D. Issatchenkia | Ans.-A

10. Which is known as Bakers yeast?

A. Saccharomyces cerevisiae | B. Candida
C. Pichia | D. Issatchenkia | Ans.-A

11. Probiotics contains a group of which bacteria?

A. *Lactobacillus* sp. | B. *Acetic acid bacteria*
C. *Formic acid bacteria* | D. *Citric acid bacteria* | Ans.-A

12. What is the shape of *Bifidobacterium*?

A. Forked shaped | B. Round
C. Rod | D. Oval | Ans.-A

13. Yogurt is fermented with the help of what kind of fermentation process?

A. Thermophilic | B. Mesophilic
C. Psychrophilic | D. Acidophilic | Ans.-A

14. The major substrate used for the *Haria* preparation?

A. Glutinous rice | B. Boiled rice
C. Whaet | D. Millet | Ans.-A

15. *Villi* fermentation done with the help of?

A. Moulds | B. Yeasts
C. Bacteria | D. Actinomycetes | Ans.-A

16. *Idli* is fermented with the help of what kind of microbes?

A. Moulds | B. Yeasts
C. Lactic acid Bacteria | D. Actinomycetes | Ans.-C

17. *Palm wine* fermentation done by?

A. Moulds | B. Yeasts
C. Lactic acid Bacteria | D. Actinomycetes | Ans.-B

18. *Haria* a (rice beer) fermentation done by?
A. Moulds
B. Yeasts
C. Lactic acid Bacteria
D. Actinomycetes
Ans.-B

19. *Saurkraut* is fermented with which bacteria?
A. *Leuconostoc mesenteroides*
B. *Saccharomyces cerevisiae*
C. *Candia*
D. *Pichia*
Ans.-A

20. *Kimchi* is a national formented food from which country?
A. Korea
B. South Africa
C. America
D. Indonesia
Ans.-A

21. Sake is national drink of which country?
A. Japan
B. South Africa
C. America
D. Indonesia
Ans.-A

22. Amylolytic Starter cultures have aboundsance of which microbes?
A. Yeasts
B. Moulds
C. Lactic acid Bacteria
D. Actinomycetes
Ans.-A

23. Kefir is a?
A. fermented milk product
B. fermented soya product
C. fermented rice beer
D. fermented disttiled product.
Ans.-A

24. The role of Probiotics is?
A. Cholestral lowering property
B. Anticancerous property
C. Anti-tumorous property
D. All of the above
Ans.-D

24. The Roquefort cheese is made by?
A. *Penicillium roquefortii*
B. *Penicillium notatum*
C. *Penicillium chrysogenum*
D. All of the above
Ans.-A

25. The Swees cheese is made by?
A. *Lactobacillus bulgaricus and Streptococcus thermophilus*
B. *Penicillium notatum*
C. *Penicillium chrysogenum*
D. All of the above
Ans.-A

26. The Camembber cheese is made by?
A. *Penicillium camembertii*
B. *Penicillium notatum*
C. *Penicillium chrysogenum*
D. All of the above
Ans.-A

27. Acidophillus milk is produced by?
A. *Lactobacillus acidophilus* B. *Penicillium notatum*
C. *Penicillium chrysogenum* D. All of the above Ans.-A

28. Bulgarian buttermilk milk is produced by?
A. *Lactobacillus bulgaricus* B. *Penicillium notatum*
C. *Penicillium chrysogenum* D. All of the above Ans.-A

29. Kefir is produced by?
A. *Lactobacillus bulgaricus* and *Streptococcus lactis* B. *Penicillium notatum*
C. *Penicillium chrysogenum* D. All of the above Ans.-A

30. Kefir is produced by which yeast?
A. *Saccharomyces delbrueckii* B. *Penicillium notatum*
C. *Penicillium chrysogenum* D. All of the above Ans.-A

31. Phosphatase test done for?
A. To checks whether milk is adequately pasteurized. B. Sterilised or not
C. Contamination of milk D. All of the above Ans.-A

32. Pasteurization of milk done at?
A. 62.8° for 30 minutes B. 69.8° for 30 minutes
C. 68.8° for 30 minutes D. All of the above Ans.-A

33. Pasteurization by HTST done at?
A. 71.7° for 15 seconds B. 69.8° for 30 minutes
C. 68.8° for 30 minutes D. All of the above Ans.-A

34. Pasteurization kills?
A. *Coxiella burnetii* B. *Ecoli*
C. *Bacillus* D. All of the above Ans.-A

35. Pasteurization was discovered by?
A. *Louis Pasteur* B. *Waksmann*
C. *Rchard petrii* D. All of the above Ans.-A

36. Acid producers in milk are?
A. *Streptococcus lactis and S.cremoris* B. *Bacillus*
C. *E.coli* D. All of the above Ans.-A

37. Gas producers in milk are?

A. *Coliforms, Clostrridium. butyricum* B. *Bacillus*

C. *Pichia* D. All of the above Ans.-A

38. Ropiness of milk is caused by?

A. *Alkaligens viscolactis* B. *Bacillus*

C. *Pichia* D. All of the above Ans.-A

39. Proteolytic bacteria in milk are?

A. *Bacillus subtilis, Bacillus.cereus* B. *Micrococcus*

C. *Pichia* D. All of the above Ans.-A

40. Lipolytic bacteria in milk are?

A. *Pseudomonas fluoresciens, Candida lipolytica* B. *Micrococcus*

C. *Pichia* D. All of the above Ans.-A

41. Homofermentative bacteria in milk are?

A. *Lactobacillus Acidophilus* B. *Micrococcus*

C. *Pichia* D. All of the above Ans.-A

42. Meat spoilage caused by?

A. *Micrococcus, Achromobacter, Pseudomonas* B. *E.coli*

C. *Pichia* D. All of the above Ans.-A

43. Poultry spoilage caused by?

A. *Achromobacter, Pseudomonas, Salmonella* B. *E.coli*

C. *Pichia* D. All of the above Ans.-A

44. Fish spoilage caused by?

A. *Achromobacter, Pseudomonas, Vibrio* B. *E.coli*

C. *Pichia* D. All of the above Ans.-A

45. Egg spoilage caused by?

A. *Pseudomonas, penicillium* B. *E.coli*

C. *Pichia* D. All of the above Ans.-A

46. Sausage spoilage caused by?

A. *Micrococcus, Lactobacillus, Streptococcus* B. *E.coli*

C. *Pichia* D. All of the above Ans.-A

47. Bread mould?
A. *Rhizopous stolonifer* B. *E.coli*
C. *Pichia* D. All of the above Ans.-A

48. Ropiness of bread is caused by?
A. *Bacillus subtilis* B. *E.coli*
C. *Pichia* D. All of the above Ans.-A

49. Ropiness of bread is caused by?
A. *Bacillus subtilis* B. *E.coli*
C. *Pichia* D. All of the above Ans.-A

50. The red Bread mould is?
A. *Monilia* B. *E.coli*
C. *Pichia* D. All of the above Ans.-A

51. In canned foods the spoilage organisms are reffered to as?
A. *Sulhide stinkers* B. *Sulpherstinkers*
C. *Zink stinkers* D. All of the above Ans.-A

52. In canned foods the endospore former is?
A. *Bcillus megaterium* B. *Sulpherstinkers*
C. *Zink stinkers* D. All of the above Ans.-A

53. In Saurkraut which substrate is used?
A. *Cabbage* B. *Pottato*
C. *Tomato* D. Beans Ans.-A

54. Rennet is produced from which microbe?
A. *Mucor* B. *Aapergillus*
C. Pichia D. Candida Ans.-A

55. Rennet is produced from which microbe?
A. *Mucor* B. *Aapergillus*
C. Pichia D. Candida Ans.-A

56. Soft pickle is produced from which microbe?
A. *Penicillum* B. *Aapergillus*
C. Pichia D. Candida Ans.-A

57. Green olives produced from which microbe?
A. *Leuconostoc mesenteroides* B. *Aapergillus*
C. Pichia D. Candida Ans.-A

58. Spoilage of olives caused due to which microbe?

A. *Enterobacter* spp. B. *Aapergillus*

C. Pichia D. Candida Ans.-A

59. Bacterial soft rot is caused due to?

A. *Erwinia carotovara* B. *Aapergillus*

C. Pichia D. Candida Ans.-A

60. Grey mould soft rot is caused by?

A. *Botrytis* spp. B. *Aapergillus*

C. Pichia D. Candida Ans.-A

61. Blue mould rot is caused by?

A. *Penicillium* spp. B. *Aapergillus*

C. Pichia D. Candida Ans.-A

62. Powdery mildew is caused by?

A. *Phytophthora* spp. B. *Aapergillus*

C. Pichia D. Candida Ans.-A

63. Black mould rot is caused by?

A. *Aspergillus spp* B. *Bacillus*

C. Pichia D. Candida Ans.-A

64. Black rot is caused by?

A. *Alternaria* spp. B. *Bacillus*

C. Pichia D. Candida Ans.-A

65. Pink rot is caused by is caused by?

A. *Trichothecium* B. *Bacillus*

C. Pichia D. Candida Ans.-A

66. *Fusarium* rot is caused by is caused by?

A. *Fusarium* spp. B. *Bacillus*

C. Pichia D. Candida Ans.-A

67. *Green mould* rot is caused by is caused by?

A. *Cladosporium* B. *Bacillus*

C. Cichia D. Candida Ans.-A

68. *Brown* rot is caused by is caused by?

A. *Sclerotinia* B. *Bacillus*

C. Pichia D. Candida Ans.-A

69. *Watery soft* rot is caused by is caused by?

A. *Sclerotinia* B. *Bacillus*

C. Pichia D. Candida Ans.-A

70. Spoilage of fruit and fruit juices is caused by is caused by?

A. *Leuconostoc mesentroides* B. *Bacillus*

C. Pichia D. Candida Ans.-A

71. Putrefaction is caused by?

A. *Clostridium putrefaciens* B. *Bacillus*

C. Pichia D. Candida Ans.-A

72. Red spot on meat is caused by?

A. *Serretia* spp. B. *Bacillus*

C. Pichia D. Candida Ans.-A

73. Greenish spot on Sausages is caused by?

A. *Leuconostoc* B. *Bacillus*

C. Pichia D. Candida Ans.-A

74. Greenish spot on Sausages is caused by?

A. *Leuconostoc* spp. B. *Bacillus*

C. Pichia D. Candida Ans.-A

75. Spoilage of canned food is caused by?

A. *Bacillus coagulans,* *B. thermosaccharolyticum* B. *Bacillus*

C. Pichia D. Candida Ans.-A

76. Sulphur stinker is caused by?

A. *Desulfotomaculum nigrificans* B. *Bacillus*

C. Pichia D. Candida Ans.-A

77. Black beets are caused by?

A. *Bacillus betanigrificans* B. *Bacillus*

C. Pichia D. Candida Ans.-A

78. Spoilage of canned meat is caused by?

A. *Bacillus subtilis* B. *E coli*

C. Pichia D. Candida Ans.-A

79. Spoilage of flour is caused by?

A. *Acetobacter* spp. B. *E coli*

C. Pichia D. Candida Ans.-A

80. Spoilage of honey is caused by?

A. *Zygosaccharomyces* spp. B. *E coli*

C. Pichia D. Candida Ans.-A

81. Spoilage of syrup is caused by?

A. *Eenterobacter aerogens* B. *E coli*

C. Pichia D. Candida Ans.-A

82. Spoilage of sugar is caused by?

A. *Saccharomyces* spp. B. *E coli*

C. Pichia D. Candida Ans.-A

83. Spoilage of apple juice is caused by?

A. *Lactobacillus pastorianus* B. *E coli*

C. Pichia D. Candida Ans.-A

84. Spoilage of pear juice is caused by?

A. *Lactobacillus plantrum* B. *E coli*

C. Pichia D. Candida Ans.-A

85. Spoilage of grape juice is caused by?

A. *Streptococci* B. *E coli*

C. Pichia D. Candida Ans.-A

86. Spoilage of orange juice is caused by?

A. *Lactobacillus pastorianus* B. *E coli*

C. Pichia D. Candida Ans.-A

87. The top yeast is?

A. *Saccharomyces cerevisiae* B. *Issatchenkia*

C. Pichia D. Candida Ans.-A

88. The bottom yeast is?

A. *Saccharomyces cerevisiae* B. *Saccharomyces uvarum*

C. Pichia D. Candida Ans.-B

89. The coffee fermenting flora is?

A. *Erwinia dissolvens* B. *Issatchenkia*

C. Pichia D. Candida Ans.-A

90. The Cocoa fermenting flora is?

A. *Candida crusei* B. *Issatchenkia*

C. Pichia D. Candida Ans.-A

91. The Garri fermenting flora is?
A. *Geotricum* spp B. *Issatchenkia*
C. Pichia D. Candida Ans.-A

92. The Tempeh fermenting flora is?
A. *Rhizopus oligosporus* B. *Issatchenkia*
C. Pichia D. Candida Ans.-A

93. The fermenting flora of Sausage is?
A. *Pediococcus cerevisiae* B. *Issatchenkia*
C. Pichia D. Candida Ans.-A

94. The film yeast is?
A. *Pediococcus cerevisiae* B. *Issatchenkia*
C. Pichia,Candida D. Rhizopus oligosporus Ans.-C

95. The beer disease is caused by?
A. *S.diastaticus* B. *Issatchenkia*
C. Pichia,Candida D. Rhizopus oligosporus Ans.-A

96. The beer disease is caused by which bacteria?
A. *Clostridium, Bacilus* B. *Issatchenkia*
C. Pichia,Candida D. Rhizopus oligosporus
Ans.-A

97. Wishkers on meat caused by which microbe?
A *Thamnidium elegans* B. *Issatchenkia*
C. Pichia,Candida D. Rhizopus oligosporus Ans.-A

98. Which mould is known as dairy mould?
A. *Geotricum candidum* B. *Issatchenkia*
C. Pichia, Candida D. Rhizopus oligosporus Ans.-A

99. Which mould is known as pink mould?
A. *Trichothecium* B. *Issatchenkia*
C. Pichia,Candida D. Rhizopus oligosporus Ans.-A

100. Which mould is known as red bread mould?
A. *Neurospora* B. *Issatchenkia*
C. Pichia,Candida D. Rhizopus oligosporus Ans.-A

101. Which mushrooms used as direct food source?
A. *Agaricus bisporus* B. *Issatchenkia*
C. Pichia, Candida D. Rhizopus oligosporus Ans.-A

4

Industrial Microbiology

1. Which one of the following is called as the 'brewer's yeast'

a) *Aaccharomyces ludwigi* b) *Saccharomyces cerevisiae*

c) *Saccharomyces boulardii* d) *Saccharomyces pastorianus*

Ans.-B

2. Butanol is obtained by fermenting molasses by :

a) *Clostridium butyricum* and *Clostridium acetobutylicum*

b) *Clostridium butyricum and Clostridium tetanai*

c) *Clostridium butyricum and lactobacillus*

d) *Clostridium butyricum and Clostridium oceanicum* Ans.-A

3. Which of the following is the source of Vitamin A

a) Sterptococcus b) *Rhodotorula gracilis*

c) Both a and b d) Yeast Ans.-B

4. Yeast cells are good source of

a) Vitamin A and B D

c) Vitamin B and D

d) all of these Ans.-C

5. Roquefort cheese is ripened by

a) *Penicillium roqueforti* b) *Penicillium camemberti*

c) All *Penicillium sp* d) *Penicillium rhizogenes*

Ans.-A

6. *Penicillium camemberti is used for ripening of*

a) Roqueforti cheese b) Camembert cheese

c) All cheese d) Fruits Ans.-B

7. Unicelled microbes grown as source of proteins are called :
   a) Microbial proteins b) Single cell proteins
   c) Unicelled proteins d) None of these Ans.-B
8. Which of the following are rich source of protein?
   a) *Spirulina and Chlorella* b) *Chlorella and Scendesmus*
   c) *Scenedesmus* d) all of the above Ans.-D
9. Zymase is obtained from
   a) *Saccharomyces ludwigi* b) *Saccharomyces cerevisiae*
   c) *Saccharomyces ludwigi* d) *Saccharomyces boulardii* Ans.-B
10. Pectinase is obtained from
   a) *Saccharomyces cerevisiae* b) *Aspergillus* sp
   c) *Spirulina* d) *Penicillium* sp Ans.-D
11. Glucose oxidase is obtained from
   a) *Saccharomyces cerevisiae* b) *Aspergillus niger*
   c) *Spirulina* d) *Penicillium* sp Ans.-B
12. *Bacillus thuringiensis* is used as
   a) Insecticide b) Fungicide
   c) Microbicidal agent d) Rodenticide Ans.-A
13. Petroleum degrading species include
   a) *Nocardia* b) *Micrococcus*
   c) *Penicillium* d) All of the above Ans.-D
14. Kojic acid is obtained from
   a) *Nocardia* b) *Micrococcus*
   c) *Penicillium* d) Aspergillus Ans.-D
15. Fumaric acid is obtained from
   a) *Saccharomyces cerevisiae* b) *Aspergillus niger*
   c) *Rhizopus stolonifer* d) *Penicillium* sp Ans.-C
17. The symbiotic relationship between fungi and roots of higher plants is called
   a) Lichen b) Mycorrhiza
   c) Helotism d) Mutualism Ans.- B
18. The advantage of fungus in this association is :
   a) Food b) Protection
   c) Mineral absorption d) All of these Ans.-A

19. In mycorrhiza, the fungus may form colonies :
a) Extracellularly b) Intracellularly
c) Both a and b d) Depends on conditions Ans.-C

20. The advantage of plants in this association is :
a) Food
b) Protection
c) Increased mineral absorption and disease protection
d) All of these Ans.-C

21. The endomycorrhizas are also called as :
a) Hartig nets
b) Mat forming mycorrhizas
c) Vesicular Arbuscular Mycorrhiza (VAM)
d) Intracellular mycorrhizas Ans.-C

22. The ectomycorrhizas are commonly formed in
a) Herbaceous plants b) Woody plants
c) All plants d) Grasses Ans.-B

23. The ectomycorrhizas form an intercellular network in root cortex called
a) Arbuscules b) Vescicles
c) Hartig net d) Haustoria Ans.-C

24. The characteristic feature of VAM is it penetrates plant cell wall and form
a) Spores intracellularly
b) Vescicles and dichotomously branched invaginations called arbuscules
c) Haustoria
d) Massive spore forming structures intracellularly Ans.-B

25. The major advantage of a plant with VAM is
a) Increased N2 absorption b) Increased P absorption
c) Increased K absorption d) Increased Mn absorption
Ans.-B

26. The fungal partner in ectomycorrhiza belongs to the class :
a) Basidiomycetes b) Ascomycetes
c) Zygomycetes d) *All of the above* Ans.-D

27. The fungal partner in VAM belongs to the class
a) Basidiomycetes b) Ascomycetes
c) Zygomycetes d) *Glomeromycetes* Ans.- D

28. The endomycorrhizal association is present in
a) 10% of plant families b) 40% of plant families
c) 85% of plant families d) Less than 5% of plant families
Ans.-C

29. Heterokaryosis (presence of many genetically different nuclei in an individual) is a character noticed in
a) Endomycorrhizal fungi b) Ectomycorrhizal fungi
c) Plants roots d) Herbaceous plant roots
Ans.-A

30. The ectomycorrhizae are commonly found in :
a) 10% of plant families b) 40% of plant families
c) 85% of plant families d) Less than 5% of plant families
Ans.- A

31. The endomycorrhizae are found in
a) Herbaceous plants b) Woody plants
c) Grasses d) All type of plants Ans.-D

32. Agar-Agar is obtained from
a) *Gelidium* b) *Polysiphonia*
c) *Fucus* d) *Laminaria* Ans.-A

33. Plants which are not differentiated into roots, stem and leaves are grouped under
a) Gymnosperms b) Pteridophytes
c) Thallophytes d) Spermatophytes Ans.-A

34. Which are the most primitive group of algae?
a) Blue green algae b) Red algae
c) Brown algae d) Green algae Ans.-C

35. Iodine is obtained from
a) *Ulothrix* b) *Ectocarpus*
c) *Laminaria* d) *Oedogonium* Ans.-C

36. Which of the following is the most advanced group of algae
a) Cyanophyta b) Rhodophyta
c) Phaeophyta d) Chlorophyta Ans.-A

37. Which of the algae is responsible for red colour of red sea
a) *Chlamydomonas brauii* b) *Trichodesmium erythrium*
c) *Ulothrix zonata* d) None of the above Ans.-C

38. One of the following is present in blue green algae
a) Starch b) Cyanophacean granule
c) Any polysaccharide d) Floridian starch Ans.-B

39. Ability to fix atmospheric nitrogen is found in
a) Leaves of some crop plants b) Chlorella
c) Some marine red algae d) Some blue green algae Ans.-D

40. Origin and evolution of sex in algae is best seen in
a) Blue green algae b) Green algae
c) Red algae d) Brown algae Ans.-C

41. Kelps is obtained from
a) Algae b) Marine algae
c) Aquatic algae d) Lichens Ans.-A

42. Algae differ from *Riccia* ana *Marchantia* in having
a) Multicellular body b) Multicellular sex organs
c) Pyrenoids in the cell d) Thalloid body Ans.-A

43. Heterocysts are Heterocyst in *Anaebaena*
a) Green and thin walled b) Green and thick walled
c) Colourless and thin walled d) Colourless and thick walled Ans.-C

44. Zygotic meiosis is a characteristic feature of
a) *Algae* b) *Bryophytes*
c) Pteridophytes d) Gymnosperms Ans.-B

45. *Cephaleoures* is
a) An epiphytic green algae b) A parasitic green algae
c) A fresh water green algae d) A colourless red algae Ans.-B

46. Sargasso sea is named after an algae *Sargassum* which is a
a) Green algae b) Brown algae
c) Red algae d) Blue green algae Ans.-A

47. The bacteria fix nitrogen except
a) *Rhizobium* b) E.coli
c) Azotobacter d) Cyanobacteria Ans.-B

48. The iodine used in Gram staining serves as a

a) Chelator b) Catalyst

c) Mordant d) Cofactor Ans.-C

49. Which one of the following microbes used for the industrial glycerol production?

A) *Yeast* B) *Mould*

C) Actinomycetes D) *Homophiles* Ans.-A

50. Which one of the following microbes used for the industrial citrate production?

A) *Yeast* B) *Mould (Aspergillus.niger)*

C) Actinomycetes D) *Homophiles* Ans.-B

51. Which one of the following microbes used for the industrial glutamic acid production?

A) *Corynebacterium glutamicum* B) *Mould(Aspergillus.niger)*

C) Actinomycetes D) *Homophiles* Ans.-A

52. Which one of the following microbes used for the industrial ethanol production?

A) *Sacchraromyces cerevisiae* B) *Mould (Aspergillus.niger)*

C) Actinomycetes D) *Homophiles* Ans.- A

53. Which one of the following microbes used for the industrial L-lysine production?

A) *Corynebacterium glutamicum* B) *MouldI (Aspergillus.niger)*

C) Actinomycetes D) *Homophiles* Ans.-A

54. Which one of the following microbes used for the industrial Phenylalanine production?

A) *Corynebacterium glutamicum* B) *Mould (Aspergillus.niger)*

C) Actinomycetes D) *Homophiles* Ans.-A

55. Which one of the following microbes used for the industrial Tryptophane production?

A) *Corynebacterium glutamicum* B) *Mould (Aspergillus.niger)*

C) Actinomycetes D) *Homophiles* Ans.-A

56. Which scientist first identified the glutamate?

A) *Ikeda* B) *Pasteur*

C) Richard petri D) None of all Ans.-A

57. Which one of the following microbes used for the industrial vitamin B_{12} production?
A) *Propionicbacterium. shermanii* B) *Mould (Aspergillus.niger)*
C) Actinomycetes D) *Homophiles* Ans.-A

58. What is another famous name of vitamin B_{12}?
A) *Cyanocobalamine* B) *Cobalt*
C) Cobalamine D) *Homophiles* Ans.-A

59. What is major component used in production of vitamin B_{12}?
A) *Cobalt* B) *Cobalt*
C) Cobalamine D) *Homophiles* Ans.-A

60. Which one of the following microbes used for the industrial Riboflavin production?
A) *Ashbya gossypii* B) *Mould (Aspergillus.niger)*
C) Actinomycetes D) *Homophiles* Ans.-A

61. What is the name of Riboflavin production?
A) Vitamin B_2 B) Vitamin B_{12}
C) Actinomycetes D) *Homophiles* Ans.-A

62. Which one of the following microbes used for the industrial beta-carotein production?
A) *Blakeshlca trispora* B) *Mould (Aspergillus.niger)*
C) Actinomycetes D) *Homophiles* Ans.-A

63. What is the name of beta-carotein production?
A) Vitamin A B) Vitamin B_{12}
C) Actinomycetes D) *Homophiles* Ans.-A

64. Which one of the following microbes used for the industrial Lycopene production?
A) *Blakeshlea trispora* B) *Mould (Aspergillus.niger)*
C) Actinomycetes D) *Homophiles* Ans.-A

65. Which one of the following microbes used for the industrial Zeaxthin production?
A) *Flavobacterium* B) *Mould (Aspergillus.niger)*
C) Actinomycetes D) *Homophiles* Ans.-A

66. Which one of the following microbes used as source of vitamin production?
A) *Yeast* B) *Mould (Aspergillus.niger)*
C) Actinomycetes D) *Homophiles* Ans.-A

67. Which one of the following microbes used for the industrial Xanthan production?
A) *Xanthomonas. compestris* B) *Mould (Aspergillus.niger)*
C) Actinomycetes D) *Homophiles* Ans.-A

68. Which one of the following microbes used for the industrial dextran production?
A) *Leuconostoc mesenteroides* B) *MouldI(Aspergillus.niger)*
C) Actinomycetes D) *Homophiles* Ans.-A

69. Which one of the following microbes used for the industrial Alginate production?
A) *Pseudomonas aruginosa* B) *Mould (Aspergillus.niger)*
C) Actinomycetes D) *Homophiles* Ans.-A

70. Which one of the following microbes used for the industrial Scleroglucan production?
A) *Sclerotium glucanicum* B) *Mould*
C) Actinomycetes D) *Homophiles* Ans.-A

71. Which one of the following microbes used for the industrial Gellan production?
A) *Pseudomonas elodea* B) *Mould (Aspergillus.niger)*
C) Actinomycetes D) *Homophiles* Ans.-A

72. Which one of the following microbes used for the industrial Pullulan production?
A) *Aerobasidium pullulans* B) *Mould*
C) Actinomycetes D) *Homophiles* Ans.-A

73. Which one of the following microbes used for the industrial Curdlan production?
A) *Alakligens faecalis* B) *Mould (Aspergillus.niger)*
C) Actinomycetes D) *Homophiles* Ans.- A

74. Which one of the following microbes used for the industrial Emulsan production?
A) *Acinetobaacter calcoaceticus* B) *Mould (Aspergillus.niger)*
C) Actinomycetes D) *Homophiles* Ans.-A

75. Which one of the following microbes used for the industrial polyesters production?
A) *Pseudomonas oleovorans* B) *Mould (Aspergillus.niger)*
C) Actinomycetes D) *Homophiles* Ans.-A

76. Which one of the following microbes used for the industrial alpha-amylase production?

A) *Aspergillus oryzae* B) *Mould (Aspergillus.niger)*

C) Actinomycetes D) *Homophiles* Ans.-A

77. Which one of the following microbes used for the industrial amyloglucosidase production?

A) *Aspergillus niger* B) *Mould (Aspergillus.niger)*

C) Actinomycetes D) *Homophiles* Ans.-A

78. Which one of the following microbes used for the industrial cellulases production?

A) *Aspergillus niger & T.coninji* B) *Mould (Aspergillus.niger)*

C) Actinomycetes D) *Homophiles* Ans.- A

79. Which one of the following microbes used for the industrial cellulases production?

A) *Aspergillus niger & T.coninji* B) *Mould (Aspergillus.niger)*

C) Actinomycetes D) *Homophiles* Ans.-A

80. Which one of the following microbes used for the industrial glucoamylase production?

A) *Aspergillus niger* B) *Mould (Aspergillus.niger)*

C) Actinomycetes D) *Homophiles* Ans.-A

81. Which one of the following microbes used for the industrial glucose-isomerase production?

A) *Arthrobacter* spp & *Bacillus* spp.

B) *Mould (Aspergillus.niger)*

C) Actinomycetes

D) *Homophiles* Ans.-A

82. Which one of the following microbes used for the industrial invertase production?

A) *Saccharomyces cerevisiae* B) *Mould (Aspergillus.niger)*

C) Actinomycetes D) *Homophiles* Ans.-A

83. Which one of the following microbes used for the industrial keratinase production?

A) *Streptococcus fradiae* B) *Mould (Aspergillus.niger)*

C) Actinomycetes D) *Homophiles* Ans.-A

84. Which one of the following microbes used for the industrial lactase production?

A) *Saccharomyces fradiae* B) *Mould (Aspergillus.niger)*
C) Actinomycetes D) *Homophiles* Ans.- A

85. Which one of the following microbes used for the industrial lipase production?

A) *Candida lypolytica* B) *Mould (Aspergillus.niger)*
C) Actinomycetes D) *Homophiles* Ans.-A

86. Which one of the following microbes used for the industrial pectinase production?

A) *Aspergillus spp.* B) *Mould (Aspergillus.niger)*
C) Actinomycetes D) *Homophiles* Ans.-A

87. Which one of the following microbes used for the industrial penicillin acylase production?

A) *Escherichia coli* B) *Mould (Aspergillus.niger)*
C) Actinomycetes D) *Homophiles* Ans.-A

88. Which one of the following microbes used for the industrial penicilinase production?

A) *Bacillus subtilis* B) *Mould (Aspergillus.niger)*
C) Actinomycetes D) *Homophiles* Ans.-A

89. Which one of the following microbes used for the industrial protease acid production?

A) *Aspergillus niger* B) *Mould (Aspergillus.niger)*
C) Actinomycetes D) *Homophiles* Ans.-A

90. Which one of the following microbes used for the industrial pollulanase production?

A) *Klebsiella aerogens* B) *Mould (Aspergillus.niger)*
C) Actinomycetes D) *Homophiles* Ans.-A

91. Which one of the following microbes used for the industrial Takadiastase production?

A) *Aspergillus niger* B) *Mould (Aspergillus.niger)*
C) Actinomycetes D) *Homophiles* Ans.-A

92. Which one of the following microbes used for the industrial acetic acid production?

A) *Acetobacter aceti* B) *Mould (Aspergillus.niger)*
C) Actinomycetes D) *Homophiles* Ans.-A

93. Which one of the following microbes used for the industrial steroids production?

A) *Rhizopous nigricans* B) *Mould (Aspergillus.niger)*
C) Actinomycetes D) *Homophiles* Ans.-A

94. Which one of the following microbes used for the industrial b-carotein production?

A) *Duneliella* B) *Mould (Aspergillus.niger)*
C) Actinomycetes D) *Homophiles* Ans.-A

95. Which one of the following microbes used for the industrial vitamin-D production?

A) *Saccharomyces* spp. B) *Mould (Aspergillus.niger)*
C) Actinomycetes D) *Homophiles* Ans.-A

96. Which one of the following microbes used for the industrial gluconic acid production?

A) *A.niger* B) *Mould (Aspergillus.niger)*
C) Actinomycetes D) *Homophiles* Ans.-A

97. Which one of the following microbes used for the industrial itaconic acid production?

A) *A.terreus* B) *Mould (Aspergillus.niger)*
C) Actinomycetes D) *Homophiles* Ans.-A

98. Which one of the following microbes used for the industrial Kojic acid production?

A) *Aspergillus flavus* B) *Mould (Aspergillus.niger)*
C) Actinomycetes D) *Homophiles* Ans.-A

99. Which one of the following microbes used for the industrial lactic acid production?

A) *Lactobacillus delbrueckii* B) *Mould (Aspergillus.niger)*
C) Actinomycetes D) *Homophiles* Ans.-A

100. Which one of the following yeast used for the industrial CSP production?

A) *Candida utilis, Saccharomyces* spp.
B) *Mould (Aspergillus.niger)*
C) Actinomycetes
D) *Homophiles* Ans.-A

101. Which one of the following microbes used for the bioleaching process production?
A) *Thiobacillus thioxidans & T. ferroxidans*
B) *Mould (Aspergillus.niger)*
C) Actinomycetes
D) *Homophiles* Ans.-A

102. Which one of the following microbes used for the bioleaching process of nickel?
A) *Aspergillus niger* B) *Mould (Aspergillus.niger)*
C) Actinomycetes D) *Homophiles* Ans.-A

103. Which one of the following microbes used for the bioleaching process of gold?
A) *Aspergillus oryzae* B) *Mould (Aspergillus.niger)*
C) Actinomycetes D) *Homophiles* Ans.-A

104. Which bacteria is known as iron bacteria?
A) *Sphaerotilus* B) *Aspergillus.niger*
C) Actinomycetes D) *Homophiles* Ans.-A

105. Which bacteria are known as manganese reducing bacteria?
A) *Shewanelal & Geobacter* B) *Aspergillus.niger*
C) Actinomycetes D) *Homophiles* Ans.-A

106. Which bacteria are known as manganese oxidiser?
A) *Arthrobacter* B) *Aspergillus.niger*
C) Actinomycetes D) *Homophiles* Ans.-A

107. Which bacteria are known as magnetotectic bacteria?
A) *Aquaspirillum magnetotectium* B) *Aspergillus.niger*
C) Actinomycetes D) *Homophiles* Ans.-A

108. Which bacteria are known as superbug's bacteria?
A) *Pseudomonas spp.* B) *Aspergillus.niger*
C) Actinomycetes D) *Homophiles* Ans.-A

109. Which bacteria are known as iron reducing bacteria?
A) *Pseudomonas & Bacillus* B) *Aspergillus.niger*
C) Actinomycetes D) *Homophiles* Ans.-A

110. Which one of the following microbes used in BT cotton?
A) *Bacillus thuringiensis* B) *Mould (Aspergillus.niger)*
C) Actinomycetes D) *Homophiles* Ans.-A

111. Which one of the following fungi helps in phosphate solubilisation?

A) *Aspergillus & Penicillium* B) *E.coli*

C) Actinomycetes D) *Homophiles* Ans.-A

112. Which one of the following Bacteria helps in phosphate solubilisation?

A) *Pseudomonas & Bacillus* B) *E.coli*

C) Actinomycetes D) *Homophiles* Ans.-A

5

General Microbiology

1. In TEM, the microscopic column is maintained under

A. Pressure B. Vacuum

C. Temperature D. Magnetism. Ans.-B

2. Lichen is the symbiotic association of

A. Fungi & Bacteria B. Fungi & Algae

C. Algae & Bacteria D. Protozoa & Virus Ans.-B

3. In laboratory many sea bacteria grow at 300 C & 15000Ib/inch these are known as :

A. Barophilic B. Obligatory barophilic

C. Psychrophilic D. Mesophilic Ans.-A

4. In Phase contrast microscopy the special optical system makes it possible to distinguish cells which are differ slightly in their

a) Size. b) Diameter.

c) Refractive index. c) Length. Ans.-C

5. In Lichen, the fungal partner is called as :

a) Phacobiont. b) Photosymbiont.

c) Mycobiont. c) All of above. Ans.-C

6. ——————— is the type of endosymbiosis.

a) Cooperation. b) Commensalisms

c) Mutualism c) Predation Ans.-C

7. ———————is the obligatory interaction.

a) Mutualism c) Commensalism

b) Cooperation. d) Ammensalism. Ans.- A

8. The first phase contrast microscope was developed by————in 1933.
   A) Hans Janssen. B) Zacharias.
   C) Fredrick Zernike D) Lippershey Ans.-C
9. In————Microscopy the object appears dark & the microscopic field is brightly illuminated.
   a) Dark field. b) Bright field.
   c) Both c) None Ans.-B
10. Bacteria can fix the nitrogen————
   a) Pseudomonas b) Staphylococcus
   c) Rhizobium d) Lactobacillus Ans.-C
11. ————are used frequently as indices of faecal pollution in water & food.
   a) Only *E.coli*
   c) *E.coli* & *Cl.perfringens*
   d) *E.coli* & *Strept.facalis*
   d) *Cl.perfringens* & *Strept.faecalis* Ans.-A
12. *E.coli* & ————are normally referred to as fecal & non- fecal contaminants of water respectively.
   a) *C.perfringens* b) *Enterobacter aerogenes*
   c) *Bacillus* spp d) *Vibrio cholerae* Ans.-B
13. The most penetrating wavelength of UV is————
   A. 260 nm B. 230nm
   C. 248nm D. 290nm Ans.-A
14. The coli-forms bacteria are :
   A. Indicator organisms B. Beneficial organisms
   C. Gram Positive organisms D. Carcinogen Ans.-A
15. Eutrophic condition is————
   A. Nutrient Rich B. Nutrient less
   C. No nutrition D. Less Oxygenic Ans.-D
16. Oligotrophic condition is————
   A. Nutrient Less B. Nutrient Rich
   C. Microbila rich D. Fungal rich Ans.-A
17. Microorganisms fix Nitrogen with the help————
   A. Nif genes B. Tif genes
   C. R-genes D. Gal genes Ans.-A

18. Microbes that grow at low temperatue known as——————

A. Psychrophilic B. Thermophilic
C. Mesophilic D. Acidophilic Ans.-A

19. The undesirable change in food that makes it unsafe for human consumption is referred as

a) Food decay b) Food spoilage
c) Food loss d) All of the above Ans.-B

20. Food preservation involves :

a) Increasing shelf life of food
b) Ensuring safety for human consumption
c) Both a and b
d) None of these Ans.- C

21. Pasteurization is a

a) Low temperature treatment
b) Steaming treatment
c) High temperature treatment
d) low and high temperature treatment Ans.-C

22. Normally bacteria stop division

a) at 10 degree Celsius b) at 5 degree Celsius
c) at 0 degree Celsius d) at 20 degree Celsius Ans.-B

23. Which of the following statements are true about chemical preservatives

a) Microbicidal or microstatic agents
b) Chemical preservatives often hazardous to humans
c) Sodium benzoate is a widely used preservative
d) All of these Ans.-D

24. Common food poisoning microbes are

a) *Clostridium* and *Salmonella*
b) *Clostridium* and *E.coli*
c) *E.coli* and *Salmonella*
d) *Clostridium* and *Streptococcus* Ans.-A

25. Botulism is caused by :

a) *Clostridium botulinum* b) All *Clostridium* species
c) *Clostridium tetanai* d) *Clostridium subtilis* Ans.-A

26. Which of the following statements are true regarding botulinal toxin
    a) is a neurotoxin
    b) water soluble exotoxin
    c) is produced by *Clostridium botulinum,* a gram positive anaerobic bacteria
    d) all of these Ans.-C

27. Botulism prevention involves :
    a) Proper heat sterilization before food canning
    b) Addition of chemical preservatives
    c) Proper low temperature treatment before food canning
    d) All of the above Ans.-A

28. *Clostridium perfringens* poisoning is associated with :
    a) Meat products b) Vegetables
    c) Canned foods d) Fish products Ans.-A

29. *Clostridium perfingens* poison is an :
    a) Exotoxin
    b) Enterotoxin produced during sporulation
    c) Endotoxin
    d) enterotoxin produced during vegetative phase Ans.-B

30. Which of the following statements are true regarding *Staphylococcus* food poisoning
    a) is an enterotoxin
    b) causes gastroenteritis
    c) is produced by *Staphylococcus aureus*
    d) all of these Ans.-D

31. *Salmonellosis* involves
    A) An enterotoxin and exotoxin B) An enterotoxin and cytotoxin
    C) An exotoxin and cytotoxin D) A cytotoxin only Ans.-B

32. The major carrier of salmonellosis are :
    a) Meat and eggs b) Meat and fish
    c) Eggs and fish d) Eggs and fruits Ans.-A

33. Aflatoxin is produced by :
    a) *Aspergillus* sp. b) *Salmonella* sp.
    c) *Fusarium* sp. d) *Streptococcal* sp Ans.-A

34. Who is regarded as the father of microbiology?
a) Leeuwenhook b) Robert Hook
c) Louis Pasteur d) Robert Koch Ans.-C

35. Who is regarded as the father of bacteriology?
a) Leeuwenhook b) Robert Hook
c) Louis Pasteur d) Robert Koch Ans.-A

36. Who is regarded as the father of medical microbiology?
a) Leeuwenhook b) Robert Hook
c) Louis Pasteur d) Robert Koch Ans.-D

37. The term virus was coined by
a) Twort b) Iwanowsky
c) Roux d) Meyer Ans.-B

38. The term bacteriophage was coined by
a) Twort b) Iwanowsky
c) Roux d) Herelle Ans.-D

39. Germ theory of disease was postulated by
a) Edward Jenner b) Robert Hook
c) Louis Pasteur d) Robert Koch Ans.-D

40. Virus was first discovered by
a) Twort b) Iwanowsky
c) Roux d) Herelle Ans.-B

41. Bacteriophages were studied in detail by :
a) Twort b) Iwanowsky
c) Roux d) Herelle Ans.-D

42. acteria were first discovered by
a) Leeuwenhook b) Robert Hook
c) Louis Pasteur d) Robert Koch Ans.- A

43. Cyanophages were discovered by :
a) Twort and Herelle b) Bold and Tippo
c) Safferman and Morris d) Hershey and Chase Ans.- C

44. Viruses were first crystallised by :
a) Twort b) Iwanowsky
c) Roux d) Stanley Ans.-D

45. Gram staining was developed by :

a) Louis Pasteur b) Robert Koch
c) Christian Gram d) Gerald Gram Ans.-C

46. Viruses were first cultured by

a) Twort b) Iwanowsky
c) Enders d) Stanley Ans.-C

47. Cancer causing viruses were discovered by :

a) Twort b) Rous
c) Enders d) Stanley Ans.-B

48. Who is regarded as the father of virology?

a) Twort b) Iwanowsky
c) Roux d) Stanley Ans.-D

49. Gram staining was developed by

a) French microbiologist Louis Pasteur
b) Dutch lens maker Leeuwenhoek
c) Danish physician Christian Gram
d) Dutch physician Christian Gram Ans.-C

50. Gram staining is an example of

a) Acid fast stain b) Acid stain
c) Differential stain d) None of the above Ans.-C

51. Gram staining was developed in

a) 1882 b) 1883
c) 1884 d) 1885 Ans.-C

52. The most common stains used in Gram staining is

a) Crystal violet and methylene blue
b) Crystal violet and safranin
c) Crystal violet and carbol fuschin
d) Safranin and methylene blue Ans.- B

53. Which of the following statements are true regarding Gram positive bacteria

a) Cell wall has a thick peptidoglycan layer
b) Cell wall lipid content is very low
c) Lipopolysaccharide layer is absent
d) all of these Ans.-A

54. Which of the following statements is incorrect regarding Gram negative bacteria
 a) Cell wall has a thin peptidoglycan layer
 b) Cell wall lipid content is very low
 c) Lip-polysaccharide layer is present
 d) All of these Ans.-B

55. In Gram staining, if some bacteria retain the crystal violet stain after alcohol treatment. Then the bacteria is :
 a) Gram positive
 b) Gram negative
 c) Procedure is incomplete to answer this question
 d) None of these Ans.-A

56. Counter stain used in Gram staining is
 a) Safranin b) Crystal violet
 c) Carbol fuschin d) Acetocarmine Ans.-A

57. In Gram staining , the alcohol acts on
 a) Teichoic acids b) Periplasm
 c) Membrane lipids d) Peptidoglycan Ans.-C

58. Lipopolysaccharide is found in cell wall of
 a) Gram positive bacteria b) Gram negative bacteria
 c) Both d) Fungi Ans.- B

59. After ethanol treatment Gram negative bacteria can be visualised
 a) Only by counter staining with safranine
 b) In violet colour of crystal violet
 c) Only by addition of iodine solution
 d) None of these Ans.-A

60. Which of the following is a common Gram positive bacteria
 a) *Rhizhobium* of root nodules b) *Lactobacillus* in curd
 c) *Escherichia coli* d) None of these Ans.-B

61. During ethanol treatment in gram staining,
 a) Crystal violet stain is dissolved in alcohol in Garm negative bacteria due to high lipid content in the outer layer
 b) Crystal violet stain is retained in Garm positive bacteria due to less lipid content and thick peptidoglycan wall
 c) Both a and b
 d) None of these Ans.-C

62. The bacteria fix nitrogen except
    a) Rhizobium   b) E.coli
    c) Azotobacter   d) Cyanobacteria   Ans.-B
63. The iodine used in Gram staining serves as a
    a) Chelator   b) Catalyst
    c) Mordant   d) Cofactor   Ans.-C
64. Which among the following is called as filamentous bacteria?
    a) Mycoplasmas   b) Spirochetes
    c) Actinomycetes   d) Vibrios   Ans.-C
65. Which of the following group of bacteria is considered as a link between bacteria and virus?
    a) Mycoplasmas   b) Spirochaetes
    c) Actinomycetes   d) Vibrios   Ans.-A
66. Cork-screw shaped forms of bacteria are
    a) Bacilli   b) Stalked bacteria
    c) Spirochaetes   d) Actinomycetes   Ans.-C
67. The ability of bacteria to change their morphological form frequently is termed as
    a) Lysogeny   b) Pleomorphism
    c) Alteromorphism   d) None of these   Ans.-B
68. Bacterial cell wall is made up of
    a) Chitin   b) Cellulose
    c) Dextran   d) Peptidoglycan   Ans.-D
69. Bacterial flagella is made up of
    a) Microtubules   b) Tubulin
    c) Flagellin   d) Spinin   Ans.-C
70. Surface appendage of bacteria meant for cell-cell attachment during conjugation is
    a) Pili   b) Flagella
    c) Spinae   d) Cilia   Ans.-A
71. Bacterial chromosome is
    a) Single stranded and circular
    b) Double stranded and circular
    c) Single stranded and linear
    d) Double stranded and linear   Ans.-B

72. Extra chromosomal, circular, double stranded, self-replicating DNA molecule in bacteria is called
a) Cosmid b) Plasmid
c) Phagemid d) Phasmid Ans.-B

73. Who is regarded as the father of microbiology?
a) Leeuwenhook b) Robert Hook
c) Louis Pasteur d) Robert Koch Ans.- C

74. Who is regarded as the father of bacteriology?
a) Leeuwenhook b) Robert Hook
c) Louis Pasteur d) Robert Koch Ans.- A

75. Who is regarded as the father of medical microbiology?
a) Leeuwenhook b) Robert Hook
c) Louis Pasteur d) Robert Koch Ans.- D

76. The term virus was coined by
a) Twort b) Iwanowsky
c) Roux d) Meyer Ans.-B

77. The term bacteriophage was coined by
a) Twort b) Iwanowsky
c) Roux d) Herelle Ans.- D

78. Germ theory of disease was postulated by
a) Edward Jenner b) Robert Hook
c) Louis Pasteur d) Robert Koch Ans.-D

79. Virus was first discovered by
a) Twort b) Iwanowsky
c) Roux d) Herelle Ans.-B

80. Bacteriophages were studied in detail by
a) Twort b) Iwanowsky
c) Roux d) Herelle Ans.-D

81. Bacteria were first discovered by
a) Leeuwenhook b) Robert Hook
c) Louis Pasteur d) Robert Koch Ans.- A

82. Cyanophages were discovered by
a) Twort and Herelle b) Bold and Tippo
c) Safferman and Morris d) Hershey and Chase Ans.-C

83. Viruses were first crystallised by

a) Twort b) Iwanowsky

c) Roux d) Stanley Ans.-D

84. Gram staining was developed by

a) Louis Pasteur b) Robert Koch

c) Christian Gram d) Gerald Gram Ans.-C

85. Viruses were first cultured by

a) Twort b) Iwanowsky

c) Enders d) Stanley Ans.-C

86. Cancer causing viruses were discovered by :

a) Twort b) Rous

c) Enders d) Stanley Ans.-B

87. Who is regarded as the father of virology?

a) Twort b) Iwanowsky

c) Roux d) Stanley Ans.-D

88. Gram staining was developed by

a) French microbiologist Louis Pasteur

b) Dutch lens maker Leeuwenhoek

c) Danish physician Christian Gram

d) Dutch physician Christian Gram Ans.-C

89. Gram staining is an example of

a) Acid fast stain b) Acid stain

c) Differential stain d) None of the above Ans.-C

90. Gram staining was developed in

a) 1882 b) 1883

c) 1884 d) 1885 Ans.-C

91. The most common stains used in Gram staining is :

a) Crystal violet and methylene blue

b) Crystal violet and safranin

c) Crystal violet and carbol fuschin

d) Safranin and methylene blue Ans.-B

92. Which of the following statements are true regarding Gram positive bacteria

a) Cell wall has a thick peptidoglycan layer

b) Cell wall lipid content is very low

c) Lipopolysaccharide layer is absent

d) All of these

Ans.-A

93. The Counter stain used in Gram staining is

a) Safranin b) Crystal violet

c) Carbol fuschin d) Acetocarmine

Ans.- A

6

Microbial Growth and Growth Control

1. Microbial growth is best in which phase of growth curve?
   A. Lag Phase — B. Log phase
   C. Stationary phase — D. Death phase — Ans.-B
2. Microbial growth is less in which phase of growth curve?
   A. Lag Phase — B. Log phase
   C. Stationary phase — D. Death phase — Ans.-D
3. Microbial growth is stable in which phase of growth curve?
   A. Lag Phase — B. Log phase
   C. Stationary phase — D. Death phase — Ans.-C
4. Phototrophs obtained energy from?
   A. Light — B. Water
   C. Chemical — D. Rock — Ans.-A
5. Chemotrophs obtained energy from?
   A. Light — B. Water
   C. Oxidation of chemical — D. .... — Ans.-???
6. Chemotrophs obtained energy from?
   A. Light — B. Water
   C. Oxidation of chemical — D. Rock — Ans.-C
7. Organotrophs obtained energy from?
   A. Organic matters — B. Water
   C. Oxidation of chemical — D. Rock — Ans.-A
8. Litholtrophs are?
   A. Rock eaters — B. Water eaters
   C. Chemical eaters — D. Organic eaters — Ans.-A

8. Methanotrophs are?

A. Methane utiliser	B. Water eaters			
C. Chemical eaters	D. Organic eaters	Ans.-A		

9. EMB is used for the growth of?

A. E.coli	B. Bacillus	
C. Pichia	D. Candida	Ans.-A

10. MacConkey agar is used for the growth of?

A. E.coli	B. Bacillus	
C. Pichia	D. Candida	Ans.-A

11. PDA is used for the growth of?

A. Fungi.	B. Bacillus	
C. Pichia	D. Candida	Ans.-A

12. YMA is used for the growth of?

A. Yeast	B. Bacillus	
C. Pichia	D. Candida	Ans.-A

13. Blood agar is used for the growth of?

A. Streptococci	B. Bacillus	
C. Pichia	D. Candida	Ans.-A

14. Chocolate agar is used for the growth of?

A. Pneumococcus	B. Bacillus	
C. Pichia	D. Candida	Ans.-A

15. Hiss serum water medium is used for the growth of?

A. Pneumococcus	B. Bacillus	
C. Pichia	D. Candida	Ans.-A

16. Tellurite blood agar medium is used for the growth of?

A. Corynebacterium	B. Bacillus	
C. Pichia	D. Candida	Ans.-A

17. XLD medium is used for the growth of?

A. Sheigella and Salmonella	B. Bacillus	
C. Pichia	D. Candida	Ans.-A

18. SS agar medium is used for the growth of?

A. Sheigella and Salmonella	B. Bacillus	
C. Pichia	D. Candida	Ans.-A

19. Desoxycholate citrate agar medium is used for the growth of?
A. Sheigella and Salmonella B. Bacillus
C. Pichia D. Candida Ans.-A

20. Tetrathionate broth medium is used for the growth of?
A. Salmonella B. Bacillus
C. Pichia D. Candida Ans.-A

21. TCBS agar medium is used for the growth of?
A. *Vibrio* B. Bacillus
C. Pichia D. Candida Ans.-A

22. Charcoal-yeast agar medium is used for the growth of?
A. *Legionella* B. Bacillus
C. Pichia D. Candida Ans.-A

23. Amies medium is used for the growth of?
A. *Gonococci* B. Bacillus
C. Pichia D. Candida Ans.-A

24. Bordet Gengou medium is used for the growth of?
A. *Bordetella* B. Bacillus
C. Pichia D. Candida Ans.-A

25. *Baird Parker Agar* medium is used for the growth of?
A. *Staphylococci* B. Bacillus
C. Pichia D. Candida Ans.-A

26. Yeast extract mannitol agar medium is used for the growth of?
A. *Rhizobium* B. Bacillus
C. Pichia D. Candida Ans.-A

27. Luria-Bertani (LB) medium is used for the growth of?
A. *Azospirillum* B. Bacillus
C. Pichia D. *E.coli* Ans.-B

28. Cetrimide Agar medium is used for the growth of?
A. *Pseudomonas* B. Bacillus
C. Pichia D. Candida Ans.-A

29. Azotobacter agar medium is used for the growth of?
A. *Azotobacter* B. Bacillus
C. Pichia D. Candida Ans.-A

30. Acetobacter Agar medium is used for the growth of?
A. *Acetobacter* B. Bacillus
C. Pichia D. Candida Ans.-A

31. Axenic culture is?
A. Pure culture B. Contaminateed culture
C. Yeast culture D. Broth culture Ans.-A

32. The term Gnotobiotic stands for?
A. Germ free conditions B. Microbial Culture
C. Contaminated D. Cell line Ans.-A

33. Examples of eukaryotic microbes?
A. *Yeast and Fungi* B. Bacillus
C. *E.coli* D. Streptococci Ans.-A

34. Examplesendosporeforming genera of bacteria?
A. *Yeast and Fungi* B. Bacillus & Clostridium
C. *E.coli* D. Streptococci Ans.-B

35. Bacterial cells are dividing by?
A. Binnary fission B. Mitosis
C. Sexual reproduction D. Asexual reproduction Ans.-A

36. The primary metabolites are synthesized in which phase of growth curv?
A. Log phase B. Lag phase
C. Stationary phase D. Death phase Ans.-A

37. The Seconadry metabolites are synthesized in which phase of growth curv?
A. Log phase B. Lag phase
C. Stationary phase D. Death phase Ans.-B

38. The generation time of *E.coli*?
A. 0.35 hours B. 0.55 hours
C. 0.45 hours D. 0.85 hours Ans.-A

39. The generation time of *Bacillus subtilis*?
A. 0.43 hours B. 0.85 hours
C. 0.65 hours D. 0.95 hours Ans.-A

40. The generation time of *S aureus*?
A. 0.47 hours B. 0.85 hours
C. 0.65 hours D. 0.95 hours Ans.-A

41. The generation time of *Pseudomonas aeruginosa*?
    A. 0.58 hours B. 0.85 hours
    C. 0.65 hours D. 0.95 hours Ans.-A

42. The generation time of *Clostidium botulinum*?
    A. 0.58 hours B. 0.85 hours
    C. 0.65 hours D. 0.95 hours Ans.-A

43. The generation time of *Mycobacterium tuberclosis*?
    A. 12 hours B. 5 hours
    C. 65 hours D. 95 hours Ans.-A

44. The generation time of *Treponema pallidum*?
    A. 8 hours B. 9 hours
    C. 20 hours D. 33 hours Ans.-D

45. The generation time of *Giardia lamlia*?
    A. 18 hours B. 23 hours
    C. 65 hours D. 95 hours Ans.-A

46. The generation time of *Sachharomyces cerevisiae*?
    A. 2 hours B. 23 hours
    C. 65 hours D. 95 hours Ans.-A

47. The generation time of *Monilinia fructicola*?
    A. 30 hours B. 23 hours
    C. 65 hours D. 95 hours Ans.-A

48. The Viable counts reffer?
    A. Live cells B. Dead cells
    C. Mixed Cells D. Contaminated cells Ans.-A

49. The CFU stands for?
    A. Colony forming unit B. Colony formerd union
    C. Colony forming united D. Colony forming unity

50. The Total counts reffer?
    A. Both live & dead cells B. Dead cells
    C. Mixed Cells D. Contaminated cells Ans.-A

51. The bacth culture is?
    A. Closed system B. Open system
    C. Both closed and open D. Automated system Ans.-A

52. Th Continoues culture is?

A. Closed system B. Open system

C. Both closed and open D. Automated system Ans.-B

53. The Chemostate used in?

A. Continoues culture B. bacth culture

C. Both closed and open D. Automated system Ans.-A

54. The example of barophilic bacteria?

A. *Photobacterium profundum* B. *Bacillus subtilis*

C. *E coli* D. *Pichia* Ans.-A

55. The example of osmotolerant bacteria?

A. *Staphylococcus aeureus* B. *Bacillus subtilis*

C. *E coli* D. *Pichia* Ans.-A

56. The example of Halophilic bacteria?

A. *Halobacterium salinarium* B. *Bacillus subtilis*

C. *E coli* D. *Pichia* Ans.-A

57. The example of Acidophilic bacteria?

A. *Sulfolobus* B. *Bacillus subtilis*

C. *E coli* D. *Pichia* Ans.-A

58. The example of Neutrophilic bacteria?

A. *E.coli* B. *Bacillus subtilis*

C. E coli D. *Pichia* Ans.-A

59. The example of Aklkaliphilic bacteria?

A. *Bacillus alcalophilus* B. *Bacillus subtilis*

C. *E coli* D. *Pichia* Ans.-A

60. The example of facultative anaerobic bacteria?

A. *E.coli* B. *Bacillus subtilis*

C. E coli D. *Pichia* Ans.-A

61. The example of oligate anaerobic bacteria?

A. Clostridium B. *Bacillus subtilis*

C. *E coli* D. *Pichia* Ans.-A

62. The example of oligate aerobic bacteria?

A. *Micrococcus luteus* B. *Bacillus subtilis*

C. *E coli* D. *Pichia* Answer-A

63. The example of hyper-thermophilic aerobic bacteria?

A. *Sulfolobus, Pyrococcus* B. *Bacillus subtilis*

C. *E coli* D. *Pichia* Ans.-A

63. The example of Thermophilic aerobic bacteria?

A. *Thermus aquaticus* B. *Bacillus subtilis*

C. *E coli* D. *Pichia* Ans.-A

64. The example of aerotolerant anaerobic aerobic bacteria?

A. *Streptococcus pyogens* B. *Bacillus subtilis*

C. *E coli* D. *Pichia* Ans.-A

65. The water activity of pure water is?

A. 1.00 B. 2.00

C. 3.00 D. 4.00 Ans.-A

66. The water activity of bread is?

A. 0.95 B. 2.00

C. 3.00 D. 4.00 Ans.-A

67. The water activity of bread is?

A. 0.95 B. 2.00

C. 3.00 D. 4.00 Ans.-A

68. The water activity of ham is?

A. 0.90 B. 2.00

C. 3.00 D. 4.00 Ans.-A

69. The water activity of honey is?

A. 0.60 B. 2.00

C. 3.00 D. 4.00 Ans.-A

70. The water activity of chocolate is?

A. 0.60 B. 2.00

C. 3.00 D. 4.00 Ans.-A

71. The water activity of cereals is?

A. 0.70 B. 2.00

C. 3.00 D. 4.00 Ans.-A

72. The example of eukaryotic algal?

A. *Chlamydomonas nivalis* B. *Pichia*

C. Candida D. Issatchenkia Ans.-A

73. The example of fungi?
A. *Chlamydomonas nivalis* B. *Pichia*
C. *E coli* D. Bacillus Ans.-B

74. The anaerobic microorganisms?
A. *Chlamydomonas nivalis* B. *Pichia*
C. ‘Candida D. Issatchenkia Ans.-A

75. The most lethal dose of UV to kill microorganisms?
A. 260nm B. 240nm
C. 300nm D. 290nm Ans.-A

76. The most lethal dose of UV to kill microorganisms?
A. 260nm B. 240nm
C. 300nm D. 290nm Ans.-A

77. The chemical used for the cultivation of anaerobic microorganisms?
A. Sodium Thioglycolate B. Potasium Thioglycolate
C. Zinc Thioglycolate D. Mg Thioglycolate Ans.-A

78. The chemical used for the cultivation of anaerobic microorganisms?
A. Sodium Thrioglycolate B. Potasium Thioglycolate
C. Zinc Thioglycolate D. Mg Thioglycolate Ans.-A

79. The chemical used in quorum sensing in microorganisms?
A. Homoserine Lactone B. Potasium Thioglycolate
C. Zinc Thioglycolate D. Mg Thioglycolate Ans.-A

80. The alcohol used to control the bacterial growth as?
A. Disinfectant B. Antiseptic
C. Anti-metabolites D. Antibiotics Ans.-A

81. The Phenol used to control the bacterial growth as?
A. Disinfectant B. Antiseptic
C. Anti-metabolites D. Antibiotics Ans.-A

82. The Halogens used to control the bacterial growth as?
A. Antimicrobial B. Antiseptic
C. Antimetabolits D. Antibiotics Ans.-A

83. The aldehydes used to control the bacterial growth in Hospitals as?
A. Sporocidal B. Antiseptic
C. Anti-metabolites D. Antibiotics Ans.-A

84. The 70 % alcohol used to control the bacterial growth as?
A. Surface sterilents B. Antiseptic
C. Anti-metabolites D. Antibiotics Ans.-A

85. The 4 % formaldehyde used in?
A. Biological preservation B. Antiseptic
C. Anti-metabolites D. Antibiotics Ans.-A

86. The Ethylene oxide gas used for the sterilization of Petri dishes as?
A. Bactericidal and Sporo-cidal B. Antiseptic
C. Anti-metabolites D. Antibiotics Ans.-A

87. The beta propionic lactone (BPL) used for the sterilization of?
A. Vaccines and sera B. Soil
C. Utensils D. Antibiotics Ans.-A

88. Chorine is an example of?
A. Disinfectant B. Antiseptic
C. Anti-metabolites D. Antibiotics Ans.-A

89. Heavy metals are examples of?
A. Germicides B. Antiseptic
C. Anti-metabolites D. Antibiotics Ans.-A

90. The moist heat sterilization done with the help of?
A. Autoclave B. Antiseptic
C. Anti-metabolites D. Antibiotics Ans.-A

91. The dry heat sterilization done with the help of?
A. Hot oven B. Antiseptic
C. Anti-metabolites D. Antibiotics Ans.-A

92. The ionising radiation sterilization done with the help of?
A. Gamma rays and beta rays B. Antiseptic
C. Anti-metabolites D. Antibiotics Ans.-A

93. The non-ionising radiation sterilization done with the help of?
A. Gamma rays and beta rays B. UV
C. Anti-metabolites D. Antibiotics Ans.-B

94. The gamma radiation sterilization used for the?
A. Plastics and Glass wares B. Antiseptic
C. Anti-metabolites D. Antibiotics Ans.-A

95. Alcohol is good example of used for the?

A. Protein denaturants B. Antiseptic

C. Anti-metabolites D. Antibiotics Ans.-A

96. A good example for water disinfectant?

A. Chlorine B. Alcohol

C. BPL D. Antibiotics Ans.-A

97. A good example for most popular antimicrobial detergents is?

A. Quaternary ammonium compounds

B. Alcohol

C. BPL

D. Antibiotics Ans.-A

98. HEPA is used in?

A. Laminar Air Flow B. Alcohol

C. BPL D. Antibiotics Ans.-A

99. The size of HEPA is used in LAF is?

A. 0.3um B. 0.2um

C. 0.1um D. 0.6um Ans.-A

100. Antibiotics are sterilised with the help of which technique?

A. Membrane filtration B. Filtration

C. Autoclave D. Dry heat Ans.-A

101. What is the nature of bacterial cell?

A. Negatively charged B. Positively charged

C. Neutral D. Ionic Ans.-A

102. What is the nature of crystal violet?

A. Negatively charged B. Positively charged

C. Neutral D. Ionic Ans.-B

103. What is the function of alcohol in gram staining?

A. Dissolves membrane lipid B. Positively charged

C. Neutral D. Ionic Ans.-A

104. What is the function of sefranin in gram staining?

A. Counter staining B. Satining

C. Neutral D. Ionic Ans.-A

105. What is the function of sefranin in gram staining?

A. Counter staining B. Satining

C. Neutral D. Ionic Ans.-A

106. What is the function of gram staining?

A. Differentiate bacteria B. Satining

C. Neutral D. Ionic Ans.-A

107. What is the function of a stain?

A. Enhance the visibility of bacterai

B. Satining

C. Neutral

D. Ionic Ans.-A

108. What is the positively charged dye stain?

A. Crystal violet B. Sefranin

C. Orange dye D. Ionic Ans.-A

110. What is an example of acidic dye?

A. Crystal violet B. Sefranin

C. Orange dye D. Ionic Ans.-A

111. The stainig dyes absorbed by which part of bacteria?

A. Cytoplasm B. Nucleous

C. cell wall D. Plasma membrane Ans.-A

112. For the gram staining bacteria culture taken from from which phase of growth?

A. Log Phase B. Lac phase

C. Stationary phase D. Death phase Ans.-A

7

Microbial Physiology

1. The process of conversion of glucose to Pyruvate is known as?
   A. Glycolysis B. Filtration
   C. Catabolism D. Anabolism Ans.-A
2. The process of conversion of Pyruvate to glucose is known as?
   A. Gluconeogenesis B. Glycolysis
   C. Catabolism D. Anabolism Ans.-A
3. The process of conversion ammonia to nitrate is known as?
   A. Nitrification B. Glycolysis
   C. Catabolism D. Anabolism Ans.-A
4. The process of conversion of nitrate to gaseous nitrogen is known as?
   A. Denitrification B. Glycolysis
   C. Nitrification D. Anabolism Ans.-A
5. The process of conversion of nitrate to gaseous nitrogen is known as?
   A. Denitrification B. Glycolysis
   C. Nitrification D. Anabolism Ans.-A
6. The process of conversion of gaseous nitrogen to Ammonia is known as?
   A. Ammonification B. Glycolysis
   C. Nitrification D. Anabolism Ans.-A
7. The process of conversion organic matter to inorganic matters is known as?
   A. Mineralization B. Glycolysis
   C. Nitrification D. Anabolism Ans.-A

8. The process of uptake of nutrients by microbes is known as?
A. Mineralization B. Immobilization
C. Nitrification D. Anabolism Ans.-B

9. The volatile fatty acid oxidisers are?
A. *Syntrophomonas wolfeii* B. E coli
C. Bacillus D. Candida Ans.-A

10. In aerobic respiration the electron acceptor is?
A. Oxygen B. Carbon
C. Sulphur D. Nitrogen Ans.-A

11. In anaerobic respiration the electron acceptor is?
A. Carbon, Sulphur B. Potassium
C. Zink D. Oxygen Ans.-A

12. The fermentation in which only a single product is formed is known as?
A. Homolactic fermentation B. Potassium
C. Zink D. Oxygen Ans.-A

13. The fermentation in which only more than one product is formed is known as?
A. Hetero-lactic fermentation B. Potassium
C. Zink D. Oxygen Ans.-A

14. The fermentation in which different acids formed is known as?
A. Mixed acid fermentation B. Homo-lactic fermentation
C. Lactic fermentation D. fermentation Ans.-A

15. The Strickland fermentation is done by?
A. *Clostridium sporogens* B. Bacillus
C. E coli D. Issatchenkai Ans.-A

16. Ethanol fermentation done by?
A. *Yeasts* B. Bacillus
C. *E coli* D. Issatchenkai Ans.-A

17. Lactic acid fermentation done by?
A. *Lactic acid bacteria* B. Bacillus
C. E coli D. Issatchenkai Ans.-A

18. Propionic acid fermentation done by?
A. *Propionicbacterium* B. Bacillus
C. *E coli* D. Issatchenkai Ans.-A

19. Butan-di-ol fermentation done by?

A. *Clostridium* B. Bacillus
C. E coli D. Issatchenkai Ans.-A

20. 2, 3 Butan-di-ol fermentation done by?

A. *Erwinia, Bacillus* B. Pichia
C. *E coli* D. Issatchenkai Ans.-A

21. Citric acid fermentation done by?

A. *Aspergillus niger* B. Pichia
C. *E coli* D. Issatchenkai Ans.-A

22. The site of carbon-dioxide fixation is?

A. Carboxysomes B. Mitochondria
C. Nucleolus D. Mesosomes Ans.-A

23. Rhodopsin based phototrophy observed in?

A. *Halobacterium salinarium* B. *E coli*
C. *Pichia* D. *Saccharomyces* Ans.-A

24. Proteo-Rhodopsin based phototrophy observed in?

A. *Proteobacteria* B. Mitochondria
C. *Saccharomyces* D. *E coli* Ans.-A

25. Calvin cycle occurs in?

A. *Chloroplast* B. Mitochondria
C. *Nucleolus* D. *E coli* Ans.-A

26. TCA cycle occurs in which site of prokaryotes?

A. Cytoplasmic matrix B. Mitochondria
C. *Nucleolus* D. *E coli* Ans.-A

27. TCA cycle occurs in which site of eukaryotes?

A. Carboxysomes B. Mitochondrial matrix
C. *Nucleolus* D. *E coli* Ans.-B

28. ETC occurs in which site of eukaryotes?

A. Plasma membrane B. Inner mitochondrial matrix
C. *Nucleolus* D. *E coli* Ans.-B

29. ETC occurs in which site of prokaryotes?

A. Plasma membrane B. Inner mitochondrial matrix
C. *Nucleolus* D. *E coli* Ans.-A

30. Chemoosmotic model is given by?
A. Peter Mitchell B. Louis Pasteur
C. *Richard* D. Waksman Ans.-A

31. Sulpher reduction done by?
A. *Desufovibrio* B. *E coli*
C. *Bacillus* D. *Geobacillus* Ans.-A

32. Carbon reduction done by?
A. *Methanogens* B. *E coli*
C. *Bacillus* D. *Geobacillus* Ans.-A

33. Nitrate reduction done by?
A. *Pseudomonas denitrificans* B. *E coli*
C. *Bacillus* D. *Geobacillus* Ans.-A

34. Iron reduction done by?
A. *Geobacter* B. *E coli*
C. *Pichia* D. *Salmonella* Ans.-A

35. Manganese reduction done by?
A. *Shewanella* B. *E coli*
C. *Pichia* D. *Salmonella* Ans.-A

36. The photosynthctic pigments are?
A. Carotenoids & beta- carotein B. *E coli*
C. *Pichia* D. *Salmonella* Ans.-A

37. The bacterial photosynthetic pigments are?
A. Bacteriocholrophyll B. *E coli*
C. *Pichia* D. *Salmonella* Ans.-A

38. Cyanobacteria have which photosynthetic pigments?
A. Bacteriocholrophyll B. *Phycobilliproteins*
C. *Pichia* D. *Salmonella* Ans.-B

39. Photosynthetic pigments phycoerythrin absorbs light at?
A. 550 nm B. 600nm
C. 700nm D. 340nm Ans.-A

40. Photosynthetic pigments phyco-cyanin absorbs light at?
A. 620 nm B. 600nm
C. 700nm D. 340nm Ans.-A

41. Accessory Photosynthetic pigments are?
A. Carotenoids & phycobillins B. Oxygen
C. Carbon D. Photo-cell Ans.-A

42. The reaction centre of photo-system I?
A. 700nm B. 800nm
C. 600nm D. 560nm Ans.-A

43. The reaction centre of photo-system II?
A. 680nm B. 800nm
C. 600nm D. 560nm Ans.-A

44. The photosynthetic reaction centre of PNSB is?
A. 870nm B. 800nm
C. 600nm D. 560nm Ans.-A

45. The photosynthetic reaction centre of GSB is?
A. 840nm B. 800nm
C. 600nm D. 560nm Ans.-A

46. The example of Green sulphur bacteria (GSB) is?
A. *Chlorobium* B. *E coli*
C. *Bacillus* D. *Streptococcus* Ans.-A

47. The example of Green non-sulphur bacteria (GNSB) is?
A. *Chloroflexus* B. *E coli*
C. *Bacillus* D. *Streptococcus* Ans.-A

48. The example of Green non-sulphur bacteria (GNSB) is?
A. *Chloroflexus* B. *E coli*
C. *Bacillus* D. *Streptococcus* Ans.-A

49. The example of Purple-sulphur bacteria (PSB) is?
A. *Chromatium, Thiocapsa* B. *E coli*
C. *Bacillus* D. *Streptococcus* Ans.-A

50. The example of Purple-Non sulphur bacteria (PNSB) is?
A. *Rhodobacter sphaeroides* B. *E coli*
C. *Bacillus* D. *Streptococcus* Ans.-A

51. The example of purple membrane bacteria?
A. *Halobacterium salinerium* B. *E coli*
C. *Bacillus* D. *Streptococcus* Ans.-A

52. The example of bacteriorhodopsin bacteria?
   A. *Halobacterium salinerium* B. *E coli*
   C. *Bacillus* D. *Streptococcus* Ans.-A
53. *Halobacterium salinerium* is?
   A. *Archaea* B. *Bacteria*
   C. *Fungi* D. *Protozoa* Ans.-A
54. The example of ED pathway utilizing gram positive bacteria?
   A. *Halobacterium salinerium* B. *Azotobacter, Rhizobium*
   C. *Bacillus* D. *Streptococcus* Ans.-B
55. The example of ED pathway utilizing gram negative bacteria?
   A. *Halobacterium salinerium* B. *Enterococcus faecalis*
   C. *Bacillus* D. *Streptococcus* Ans.-A